43rd Northeast Bioengineering Conference (NEBEC 2017)

Newark, New Jersey, USA
31 March - 2 April 2017

ISBN: 978-1-5108-7925-6

Printed from e-media with permission by:

Curran Associates, Inc.
57 Morehouse Lane
Red Hook, NY 12571

Some format issues inherent in the e-media version may also appear in this print version.

Printed by Curran Associates, Inc. (2019)

For permission requests, please contact NJIT Department of Biomedical Engineering
at the address below.

NJIT Department of Biomedical Engineering
Fenster Hall, Room 605
University Heights
Newark, New Jersey 07102
USA

Phone: 973-596-5268

https://biomedical.njit.edu/

Additional copies of this publication are available from:

Curran Associates, Inc.
57 Morehouse Lane
Red Hook, NY 12571 USA
Phone: 845-758-0400
Fax: 845-758-2633
Email: curran@proceedings.com
Web: www.proceedings.com

Biomedical Engineering Capstone Design 2016-2017

Papers Presented at the Undergraduate Design Competition, Northeast Bioengineering Conference, NJIT, March 31, 2017

43rd NORTHEAST BIOENGINEERING CONFERENCE

March 31 - April 2

Department of Biomedical Engineering

NJIT

New Jersey Institute of Technology

ABOUT NJIT

One of the nation's leading public technological universities, New Jersey Institute of Technology (NJIT) is a top-tier research university that prepares students to become leaders in the technology-dependent economy of the 21st century. NJIT's multidisciplinary curriculum and computing-intensive approach to education provide technological proficiency, business acumen and leadership skills.

With an enrollment of 11,400 graduate and undergraduate students, NJIT offers small-campus intimacy with the resources of a major public research university. NJIT is a global leader in such fields as solar research, nanotechnology, resilient design, tissue engineering, and cybersecurity, in addition to others. NJIT is among the top U.S. polytechnic public universities in research expenditures, exceeding $130 million, and is among the top 1 percent of public colleges and universities in return on educational investment, according to PayScale.com. NJIT has a $1.74 billion annual economic impact on the State of New Jersey.

WELCOME

43rd Northeast Bioengineering Conference

New Jersey Institute of Technology
Newark, NJ

March 31 - April 2, 2017

ORGANIZING COMMITTEE

CONFERENCE CHAIR
William Hunter

FINANCE CHAIR
Xiaobo Li

CAREER EXPLORATION CHAIRS
Tara Alvarez
Bryan Pfister

RESEARCH PROGRAM COMMITTEE
Treena Arinzeh
Bharat Biswal
Eun Jung Lee
Bryan Pfister

ABSTRACT REVIEW CHAIR
Vivek Kumar

ABSTRACT REVIEWERS
Sergei Adamovich
Namas Chandra
Richard Foulds
Murat Guvendiren
James Haorah
Antje Ihlefeld
Saikat Pal
Mesut Sahin

CHECK-IN
Candida Rocha

STUDENT SUPPORT COORDINATORS
Alev Erdi
Penelope Georges

PUBLICITY
Chang Yaramothu
John D'Antonio-Bertagnolli

PROGRAM BOOKLET
Diane Cuddy
Johanna Moroch

AUDIO VISUAL
Phil Stickna

PHYSICAL PLANT
Latosha Wilson

LOCAL ARRANGEMENTS & ADMINISTRATION
Selenny Fabre

SPONSORS & EXHIBITORS

NSF
NJIT New Jersey Institute of Technology
Biomedical Engineering, Science and Health Systems
Cornell Engineering Meinig School of Biomedical Engineering
BROWN
CEMB
BINGHAMTON UNIVERSITY Thomas J. Watson School of Engineering and Applied Science
STEVENS INSTITUTE OF TECHNOLOGY
COLUMBIA BIOMEDICAL ENGINEERING
WPI

PROGRAM SCHEDULE

SCOPE OF THE CONFERENCE PROGRAM

INVITED AND SUBMITTED PRESENTATIONS

Each day begins with an invited ***Keynote Address*** by a national leader in biomedical engineering. Introductory biographies of these speakers appear on page 5 in this booklet.

Most work submitted by conference participants will be presented as ***"electronic" posters***. Some participants requested ***oral presentations***, and 40 percent of those requests were programmed into cohesive groups of talks.

On research focused days of the conference (Saturday and Sunday), faculty from the northeast region will present invited talks. One group of talks will be presented as ***Faculty Seminars***, where four invited speakers in a focused area of biomedical engineering will each introduce you to their work. In addition, each session of oral talks by participants will be introduced by an invited ***Chairman's Talk*** by the faculty chair of that platform session. Introductory biographies of invited faculty are on pages 6-8.

FINDING WORK THAT INTERESTS YOU

Titles of invited talks are indicated in the program schedule that follows in this booklet. Submitted presentations are listed in three ***xeroxed addenda*** included with this booklet: ***Design Posters, Research Posters, Oral Research Talks.*** Each list is grouped by program theme and subgrouped alphabetically by first author. The time and poster location is indicated. In addition, the NEBEC website (nebec.org) links to ***a searchable database of submitted presentations.***

FRIDAY AFTERNOON PROGRAM

1 p.m. - 1:30 p.m.	**INDUSTRY KEYNOTE SPEAKER: *Hans Hartmann*** (COO, Oculus) - **ATRIUM**		
1:45 p.m. - 3 p.m.	**BREAKOUT SESSIONS: Career Development in Biomedical Engineering**		
	BME Careers in Industry **Atrium** **Panel** *Emilio Sanchez* (Stryker) *Hans Hartmann* (Oculus) *Carla Cerqueira* (Edwards) *Amy Cilento* (Optum) *Matt Alcalde* (BioNJ)	***Clinical Careers*** **Kupfrian Seminar A** **Panel** *Karisa Solt-Schreck* (JHU) *Pooja Sheth* (Einstein Med) *Anna Barrett* (Kessler) *Saul Weiner* (NJ Dental Sch)	***Entrepreneurship*** **Kupfrian Seminar B** **Panel** *Mike Wiley* (Venture Capital) *Barry Kappel* (Sapience Therapy) *John Crombie* (UpStart Products)
3 p.m. - 5:30 p.m.	**UNDERGRADUATE DESIGN POSTERS** - **Ballroom**		
5:30 p.m. - 6 p.m.	**UNDERGRADUATE DESIGN AWARDS CEREMONY** - **Atrium**		
6 p.m. - 7:30 p.m.	**NETWORKING BUFFET WITH INDUSTRY, ACADEMIC AND CAREER LEADERS** - **Gallery**		
7:30 p.m. - 10 p.m.	**NETWORKING VIA POOL TOURNAMENT, PING-PONG AND BOWLING** - **Game Room**		

SATURDAY PROGRAM

8 a.m. - 9 a.m.	**BREAKFAST BUFFET - Gallery**
9 a.m. - 9:30 a.m.	**CONFERENCE KEYNOTE SPEAKER:** *Gilda Barabino* (President, AIMBE) **- Atrium**
9:30 a.m. - 11:30 a.m.	**RESEARCH POSTER PRESENTATIONS: Session I - Ballroom**
11:45 a.m. - 1:15 p.m.	**FACULTY SEMINAR: Biomedical Imaging - Atrium**
1:15 p.m. - 2 p.m.	**NETWORKING LUNCHEON - Gallery**
2 p.m. - 4 p.m.	**POSTER PRESENTATIONS: Session II - Ballroom**
4:15 p.m. - 5:45 p.m.	**FACULTY SEMINAR: Biomechanics & Tissue Engineering - Atrium**
6 p.m. - 7:30 p.m.	**NETWORKING BUFFET WITH RESEARCH LEADERS - Gallery**
7:30 p.m. - 10 p.m.	**NETWORKING AT THE HIGHLANDER PUB** (CASH ONLY) **- Pub**

11:45 a.m. - 1:15 p.m. — FACULTY SEMINAR: Biomedical Imaging - Atrium

Elisa Konofagou
Imaging Mechanical and Electro-mechanical Properties of the Myocardium In Vivo
Arvind Pathak
"A Pixel Is Worth a Thousand Words" - The Role of Imaging in Cancer
Andrew Tsourkas
Engineering Targeted Nanoparticles and Antibodies for Molecular Imaging
Yale Goldman
Super-resolution Optical Tracking and Structural Dynamics Molecule by Molecule

Oral Research Talks	***Oral Research Talks***	***Oral Research Talks***
Rehabilitation Engineering - *Kupfrian Seminar A* - Damiano Zanotto *Wearable Technology for Gait Training and Gait Assessments*	Tissue Engineering - *Kupfrian Seminar B* - Vivek Kumar *Moderator*	Bioengineering of Development - *Kupfrian Lecture Room* - Xiaobo Li *Moderator*

4:15 p.m. - 5:45 p.m. — FACULTY SEMINAR: Biomechanics & Tissue Engineering - Atrium

Trey Crisco
From Wrist Biomechanics, to Head Impacts, to Toy Therapies
Warren Grayson
Platform Technologies for Engineering Composite Tissues
Sanjeev Shroff
Dynamic Myocardial Contractile Parameters from Ventricular Pressure-Volume Measurements
Kevin Costa
Engineering 3-D Human Cardiac Tissues and Organoids

Oral Research Talks	***Oral Research Talks***	***Oral Research Talks***
Biomaterials - *Kupfrian Seminar A* - Sangamesh Kumbar *Micro-nanostructures for Tissue Engineering*	Biomedical Imaging - *Kupfrian Seminar B* - Bharat Biswal *Moderator*	TBI & Neural Engineering - *Kupfrian Lecture Room* - Peter Galie *Mechanical Stress Regulates Blood-brain Barrier Integrity*

PROGRAM SCHEDULE

SUNDAY MORNING PROGRAM

8 a.m. - 9 a.m.	**BREAKFAST BUFFET - Gallery**	
9 a.m. - 9:30 a.m.	**RESEARCH KEYNOTE SPEAKER: *Michele Grimm*** (National Science Foundation) **- Atrium**	
9:30 a.m. - 11:30 a.m.	**RESEARCH POSTER PRESENTATIONS: Session III - Ballroom**	
11:45 a.m. - 1:15 p.m.	**FACULTY SEMINAR: Brain Injury - Atrium**	
	Reuben Kraft *Multiscale Modeling of the Axonal Tract Level in the Brain* **Kurosh Darvish** *Nonlinear Effects in Brain Tissue Are Important in Predicting Traumatic Injury* ***Barclay Morrison*** *Primary Blast Injury Decreases Neuronal Plasticity* ***David Shreiber*** *In Situ Estimation of Axon-level Strain Following Tissue-level Stretch*	
	Oral Research Talks Tissue Engineering & Modeling *- Kupfrian Seminar A* Catherine von Reyn *Neuroengineering Strategies to Dissect and Modify Sensorimotor Circuits*	***Oral Research Talks*** Orthopaedic Biomechanics *- Kupfrian Seminar B* Deva Chan *Biomechanics of Joint Degeneration and Repair*
1:30 p.m. - 1:45 p.m.	**CLOSING CEREMONY - Atrium** Awards Announcements: Research Poster Awards and Oral Talk Awards	

EVENT LOCATIONS

The location of each event in the program is specified at the end of each line in the program schedule. All events occur either in the Campus Center or Kupfrian Hall, which is located next door to the Center. Details of each location are:

ATRIUM - First Floor of the Campus Center

BALLROOM - Second Floor of the Campus Center; enter through the Gallery

GALLERY - Second Floor of the Campus Center, outside the Ballroom

KUPFRIAN ROOMS - Lower level of Kupfrian Hall

GAME ROOM - Basement of Campus Center

PUB - Third Floor of Campus Center

KEYNOTE SPEAKERS

CONFERENCE KEYNOTE

Gilda Barabino

Dr. Barabino is dean and Berg professor at The Grove School of Engineering at The City College of New York (CCNY). She is also president of AIMBE, the most prestigious organization of professional fellows in biomedical engineering. She is also the immediate past-president of the Biomedical Engineering Society. She earned her Ph.D. from Rice University and served as a professor of BME at Georgia Tech. She is a noted investigator in the areas of sickle cell disease and cellular and tissue engineering.

INDUSTRY KEYNOTE

Hans Hartmann

Hans Hartmann is now the chief operating officer (COO) of Oculus, the leading maker of immersive virtual reality headgear. Until recently, he was the COO of Fitbit. Previously, he held a wide variety of industrial management experiences with companies that provided automotive GPS, optical networks, data storage and internet fax. He also spent 15 years at Hewlett-Packard managing quality, operations and marketing. He holds a B.S. in electrical engineering from NJIT and an M.S. in manufacturing systems engineering from Stanford.

RESEARCH KEYNOTE

Michele Grimm

Dr. Grimm is the program director of the Engineering of Biomedical Systems and Disability & Rehabilitation Engineering Programs at the National Science Foundation. Her home institution is Wayne State University, where she is a faculty member in BME. She earned her Ph.D. from the University of Pennsylvania. Her research over 25 years has engaged tissue biomechanics – from brain to bone – through which she has become a Fellow of ASME. She is an internationally recognized expert in the mechanisms of brachial plexus injury during birth.

SENIOR FACULTY TALKS

Elisa E. Konofagou is the Robert and Margaret Hariri Professor of Biomedical Engineering and Radiology at Columbia University, where she directs the Ultrasound Elasticity Imaging Lab. She earned her Ph.D. in biomedical engineering from the University of Houston and did postdoctoral work at Brigham and Women's Hospital. She is a Fellow of AIMBE. Dr. Konofagou's main interests are the development of novel elasticity imaging techniques for applications to cardiac, arterial, breast and ligament tissue as well as the blood-brain barrier.

Arvind P. Pathak is Associate Professor of Radiology, Oncology & Biomedical Engineering at Johns Hopkins. He earned his Ph.D. at the Medical College of Wisconsin/ Marquette joint BME program followed by a postdoc at Hopkins. Dr. Pathak's lab employs biomedical imaging, computational models and visualization tools to answer critical questions in cancer biology. His team investigates the micro-environment of tumors with the goal of developing clinical biomarkers and a deeper understanding of disease progression and metastasis.

Kevin D. Costa is Associate Professor of Medicine and Director of Cardiovascular Cell and Tissue Engineering at Mt. Sinai School of Medicine. He earned his Ph.D. in bioengineering at the University of California, San Diego, with postdoctoral fellowships at Johns Hopkins and Washington University. His lab focuses on multi-scale mechano-biology of cardiovascular remodeling, injury, repair and regeneration, using techniques that include atomic force microscopy, nano-medicine and 3-D tissue and organoid engineering.

Andrew Tsourkas is Professor of Bioengineering at the University of Pennsylvania, where he directs the Center for Targeted Therapeutics and Translational Nanomedicine. He earned his Ph.D. in biomedical engineering from Georgia Tech and did postdoctoral training in molecular imaging at Harvard. He is a Fellow of AIMBE. His research aims to develop new nano-formulations that can carry high payloads for therapy or imaging and develop highly efficient attachments of targeting ligands to nanoparticles.

Yale E. Goldman is Physiology Professor and former director of the Muscle Institute at the University of Pennsylvania. He obtained his MD and Ph.D. degrees at Penn and was a postdoctoral fellow at University College London. He is a Fellow and past president of the Biophysical Society. His laboratory studies molecular motors and protein synthesis using novel optical methods that include nanometer tracking of position and orientation of single fluorescent probes and ultra-fast feedback infrared optical traps.

Sanjeev Shroff is Distinguished Professor and Gerald McGinnis Chair in Bioengineering at the University of Pittsburgh. Dr. Shroff earned his Ph.D. in bioengineering from the University of Pennsylvania and is a Fellow of AIMBE. He is widely recognized for his cardiovascular research on (1) left ventricular mechano-energetic function and underlying cellular processes; (2) vascular stiffness as modified by drugs or hormones; and (3) regional contractile dyssynchrony in global ventricular mechanics and energetics.

J. J. Trey Crisco is Henry F. Lippitt Professor of Orthopedics and Professor of Engineering. He earned his Ph.D. in applied mechanics from Yale. He is Editor-in-Chief of the Journal of Applied Bio-mechanics and past president of the American Society of Biomechanics. He investigates musculoskeletal biomechanics, where he focuses on upper extremity mechanics and osteoarthritis, using advanced imaging analysis and neuromuscular therapy through toy and game play for children. He also focuses on sports performance and injury prevention.

David Shreiber is Professor of Biomedical Engineering at Rutgers. He earned his bioengineering Ph.D. at the University of Pennsylvania followed by a postdoc at the University of Minnesota. Research foci include the multi-scale analysis of CNS injury mechanics; biomaterial, tissue and cellular engineering approaches for repair and restoration of neural functions, and designing new technology for brain-machine interfaces. Recently, his lab has focused on axonal injury from stretch, and the use of chemical and mechanical gradients to guide axon growth.

Reuben Kraft is Assistant Professor of Mechanical and Biomedical Engineering at Penn State University. He obtained his Ph.D. from Johns Hopkins followed by a postdoc with the Army Research Lab. He leads a Computational Biomechanics Group that focuses on multiscale computational mechanics and methods, which are used to solve problems at the interface of biology and multiscale mechanics. He also studies blast physics and human performance in extreme environments.

Warren L. Grayson is Associate Professor of Biomedical Engineering at Johns Hopkins. He earned a biomedical engineering Ph.D. from Florida State University followed by a postdoc at Columbia. His lab addresses challenges associated with engineering functional craniofacial and orthopedic constructs for use in therapeutic applications. He develops innovative methods using stem cells to create patient-specific grafts,and designs advanced bio-reactors to maintain cell viability while precisely controlling the cellular microenvironment.

Barclay Morrison III is Professor of Biomedical Engineering at Columbia, where he leads the Neuro-trauma and Repair Lab. He earned his bioengineering Ph.D. from Penn, with a clinical neuroscience postdoc in the UK. He is associate editor of the Journal of Biomechanical Engineering and the Journal of Neurotrauma. He focuses on the biomechanics of traumatic brain injury, and the biochemical, genomic and molecular pathways responsible for post-traumatic cell death. He actively engages in educating the new crop of biomedical engineers.

Kurosh Darvish is an Associate Professor of Mechanical Engineering at Temple University. He earned his Ph.D. in mechanical and aerospace engineering and biomechanics from the University of Virginia. His research interests include soft tissue biomechanics, traumatic brain injury, traumatic aortic rupture, orthopedic devices and human body dynamics. Two of his most cited papers involve nonlinear mechanical effects in brain tissue due to viscoelasticity and large deformations.

SESSION CHAIRS

Deva Chan is Assistant Professor of Biomedical Engineering at Rensselaer Polytechnic Institute. She earned her Ph.D. from Purdue University, followed by a postdoctoral fellowship at Rush University Medical Center in Chicago. Her research aims to quantify the mechanical and biochemical mechanisms of joint degeneration and repair, and to integrate noninvasive medical imaging and computational modeling techniques to assess it.

Damiano Zanotto is Assistant Professor of Mechanical Engineering at Stevens Institute. He earned his Ph.D. in industrial engineering (emphasis on mechatronics) at the University of Padua, Italy. He was a postdoc at the University of Delaware, then became an associate research scientist in the Robotics and Rehabilitation Lab at Columbia University. He studies exoskeletons for rehabilitation and wearable sensors for motion analysis.

Peter Galie is Assistant Professor of Biomedical Engineering at Rowan University. He earned his Ph.D. from the University of Michigan. His laboratory's research focuses on how disease changes fluid flow in the brain, and how altered fluid flows affect the behavior of cells in the central nervous system. He uses microfluidic techniques to create models to control the fluid environment of cultured cells.

Xiaobo Li is Associate Professor of Biomedical Engineering and Electrical Engineering at NJIT. She earned her Ph.D. at the University of Birmingham, England, and has been a faculty member at Albert Einstein College of Medicine. She develops analytical and statistical techniques to quantitatively evaluate the structural and functional organization in the human brain using structural MRI/fMRI/DTI data.

Catherine von Reyn is Assistant Professor of Biomedical Engineering at Drexel University. She earned her Ph.D. at the University of Pennsylvania; thereafter, she was a postdoctoral associate at the Janelia Research Campus of the Howard Hughes Medical Institute. She uses genetic engineering, whole-cell patch clamp in behaving animals, modeling and detailed behavioral analysis to study sensorimotor circuits. She aims to engineer de novo synaptic connections for neurodegenerative diseases.

Vivek Kumar is Assistant Professor of Biomedical Engineering and Chemical Engineering at NJIT. He earned his Ph.D. at Georgia Tech, and then had postdoctoral stints at Rice University and Harvard Medical School. His research lies at the intersection of biomaterials, drug discovery and clinical translation, which he has applied to angiogenesis, inflammation, soft-tissue engineering and drug development.

Sangamesh Kumbar is Associate Professor of Orthopaedic Surgery and a member of the core biomedical engineering faculty at the University of Connecticut. He earned his Ph.D. at Karnatak University, India. His research aims to develop mechanically competent, porous, 3-D tissue scaffolds functionalized with nanofiber structures of natural origin for bone repair and regeneration. In addition, he aims to develop novel injectable particulate delivery systems for chemotherapeutic agents.

Bharat Biswal is Distinguished Professor of Biomedical Engineering at NJIT. He earned his Ph.D. in the biophysics imaging program at the Medical College of Wisconsin. He is internationally renowned for developing MRI techniques to study connectivity between different regions of the brain. In fact, the National Institute of Mental Health cited his research as one of the top two advancements in 2010.

The Biomedical Engineering Department at NJIT is proud to host this year's Northeast Bioengineering Conference. Just 15 years after its founding, the BME Department has blossomed into a dynamo. Fitting its role in a research-intensive university, the department excels in both missions: educate the next generation of biomedical engineers, and drive forward the frontiers of biomedical engineering knowledge. Our accredited undergraduate BME program ranks among the top 10 nationally for "best value" [1]. Our audited research expenditures exceed $3.25 million per year [2], averaging $250,000 per tenured or tenure-track faculty member. Our graduates populate numerous medical device companies in northern New Jersey: e.g., Stryker Orthopaedics, Johnson & Johnson, Becton Dickinson, Maquet, Integra Life Sciences, Edwards Lifesciences.

Welcome to BME@NJIT !

[1] http://www.bestvalueschools.com/rankings/biomedical-engineering-degrees-2016/
[2] profiles.asee.org/profiles/6951/screen/29?school_name=New+Jersey+Institute+of+Technology

Future Northeast Bioengineering Conferences

2018
Drexel University
Chair: Kenneth Barbee

2019
Rutgers University
Chair: Joseph Freeman

2020
Penn State University
Chair: Keefe Manning

PLAN TO ATTEND!!

NJIT
New Jersey Institute of Technology

University Heights
Newark, NJ 07102

500 3/17

Adjustable Lightweight Transcranial Magnetic Stimulation Helmet for Brain Injury Rehabilitation

*Rachel McAteer, Alex Gianos, Daniel Wec, Tanya Wang, Ying Sun, Ph.D., Brian Silver, MD**
Biomedical Engineering Program, University of Rhode Island, Kingston, RI 02881;
*Neurology, University of Massachusetts Medical School, Worcester, MA. Correspondence email: yingsun@uri.edu

***Abstract*— This project aims to design a lightweight helmet that utilizes the fundamentals of transcranial magnetic stimulation (TMS) to aid in the treatment of brain injuries such as stroke. The proposed design consists of a PVC framework that can hold six detachable units, two of which are motor-units for housing and rotating disc magnets. Each 3D-printed motor-unit holds two neodymium magnets, which can be rotated at two different speeds via a switch, and arranged along the framework of the helmet to target specific areas of the brain. The objective of this project is to produce a prototype helmet to support clinical studies on the efficacy of low-frequency low-intensity TMS therapy for a prolonged period of time.**

Keywords—transcranial magnetic stimulation; brain injury; rehabilitation; neodymium magnet; rotational magnet field

I. INTRODUCTION

Transcranial magnetic stimulation (TMS) is a non-invasive method of brain injury rehabilitation that uses a magnetic field to stimulate targeted areas of the brain. The treatment uses coils to create a rapidly changing magnetic field which generates weak electric currents. This causes either a hyperpolarization or depolarization of a cell's membrane potential, inducing controllable manipulations in certain behavior. Originally used for diagnostic purposes, the last few decades have seen a number of studies test the effects of TMS as a therapy option not only for psychiatric and neurological disorders [1], including depression and schizophrenia [2], but for brain injuries – specifically those attributed to strokes, such as inhibited motor control after acute cerebral infarction [3].

Current TMS therapy is performed almost exclusively in a clinical setting. A single round of treatment can take between 20 and 40 minutes, and cost about US$2,500. As the medical conditions TMS is associated with typically require frequent treatments over extended periods of time, the conventional TMS as a rehabilitation option is both costly and inconvenient. This project aims to provide an affordable, convenient, alternative to the standard TMS, in order to expand the accessibility of a promising treatment method. The proposed lightweight helmet will employ rotating permanent magnets on an adjustable framework, so that the induced magnetic field can be moved to stimulate different areas of the brain. Using a helmet, that supplies comparatively weaker stimulus, will allow patients the option of receiving treatment from home and for prolonged periods of time. While the efficacy of this approach still requires further clinical studies, the goal of the present project is to develop a safe and comfort prototype that will facilitate future clinical studies.

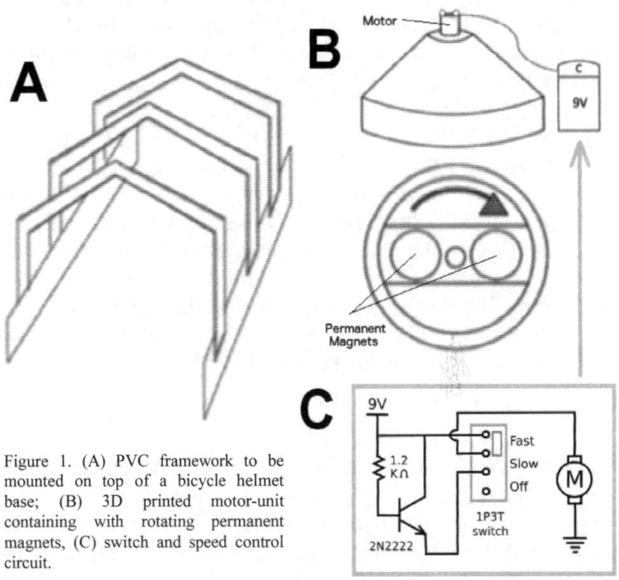

Figure 1. (A) PVC framework to be mounted on top of a bicycle helmet base; (B) 3D printed motor-unit containing with rotating permanent magnets, (C) switch and speed control circuit.

II. METHODS

The design of the TMS helmet is based on a previous project [4]. As shown in Fig. 1, the main components of this device include a framework mounted on top of a bicycle helmet base (A), two motor-units containing rotating disc magnets (B), and a switch and speed control circuit for each of the motor-units (C). These components are described as follows.

A. Framework

The framework facilitates the mounting of the motor-units and provides the flexibility of aiming the magnetic fields to specific impaired areas of the brain. The materials of the framework are from PVC pipes, cut and thermally formed to the desirable shapes. The framework contains two PVC strips that are glued to the styrofoam base of a bicycle helmet. Reclosable fasteners (3M Dual Lock® SJ3560) are used to attach three ridges across the top of the head as shown in Fig. 1A. The positions of the ridges are easily adjustable. The bottom of each ridge also has reclosable fasteners for attaching the motor-units and the support units. The support-unit has the same shape of the motor-unit, but does not contain the motor and the magnets; it is used to allow the helmet resting comfortably on the head.

B. Motor-Unit

As shown in Fig. 1B, the motor-unit consists of a small DC motor and two permanent magnets encased in a housing. The housing is designed by using a computer-aided design software (SolidWorks, Waltham, MA) and 3D-printed with ABS plastic. The motor is a generic 12V 300 rpm DC motor that has a 3 mm shaft and a speed reducer gearbox. This motor provides sufficient torque to turn the magnet assembly with the power from a 9V lithium-ion rechargeable batter (600 mAh). Each of the two disc magnets is 1/4-inch thick and 1 inch in diameter, made of N52 neodymium and having a magnetic strength of 0.33 Tesla on the surface. The support-unit is just the housing of the motor-unit and does not contain the motor and the magnets.

C. Speed Control

As shown in Fig. 1C, a simple transistor circuit is used to control the rotational speed of the magnet assemble. A single-pole triple-throw (1P3T) switch is used to select among *Off, Slow, and Fast*. The fast speed is achieved by connecting the 9V supply directly to the motor, resulting in a rotational speed of 68 rpm. The slow speed is achieved by driving the motor via an NPN transistor (2N2222). A 1200-ohm base resistor is used in order to result in a rotational speed of 34 rpm.

III. Results

The final assembly is shown in Fig. 2, which contains two motor-units and four support-units. The support-units provide stability and uniform balance so that the motor-units remain positioned correctly once the location is chosen. All units are connected to the frame by reclosable fasteners, allowing for easy rearrangement, should the area of the brain targeted for treatment change. The framework, in cooperation with the size and mobility of the motor-units, allows the helmet to target the major areas of the brain, within the cerebrum, typically injured when a stroke occurs.

Because a motor-unit operates under its own switch mechanism, each of the two motor-units implemented on a helmet at a time can operate at its own speed (either 34 rpm or 68 rpm respectively). With independently operating motors, different parts of the brain can receive different stimuli at the same time.

An important aspect in the usability of the helmet is the noise caused by the rotation of the magnets. In several tests performed in closed quiet rooms, we determined that the motors increase the decibel reading between 9 and 11 decibels on the “slow” speed. Going from the “slow” to “fast” speed results in an average increase of 10 to 12 decibels. For patients who are wearing this helmet for hours at a time, the noise at the present level may be difficult to bear.

Another issue regarding this prototype is the structural strength. Due to the strong magnetic fields, a motor-unit could be pulled out of the position by a ferromagnetic object nearby. Finally, the maximum rotational speed of the magnet assembly (68 rpm) may need to be further increased in order to enhance the intended therapeutic effects.

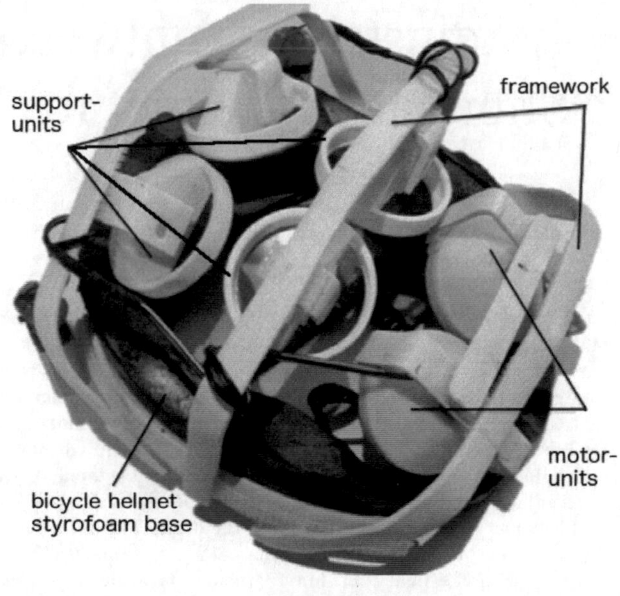

Figure 2. Functional prototype implementing two motor-units and four support-units onto PVC framework. The underside of each PVC ridge contains reclosable fasteners in order to allow all units to be rearranged as necessary.

IV. Discussion

This project has resulted in a functional prototype of the wearable TMS helmet using rotational permanent magnets. The prototype is safe and comfortable to wear on the head. However, it will need further improvements before it can be used in the clinical trial for testing the efficacy. The three issues that have been identified are noise, rotational speed, and structural strength. The motor-unit needs to be redesigned. The motor itself and the speed reduced gearbox generate too much noise. A possible choice is a servo motor cable of continuous rotation via pulse-width modulation controls. The servo motor should also be able to achieve a rotational speed of 180 rpm, about 3 times faster than what the present prototype delivers. The structure strength can be improved by increasing the contact areas of the joints. The penetration depth of the N52 magnets in use is about 10 cm based on a paper phantom. Future work will also include a penetration depth study using a more realistic phantom of the human skull.

References

[1] Hallet, M, A. Pascual-Leone, S. Rossi, and P. M. Rossini. "Safety, ethical considerations, and application guidelines for the use of transcranial magnetic stimulation in clinical practice and research." Clinical Neurophysiology, vol. 120(12), pp. 2008-2039, 2009.

[2] Paillère-Martinot et al. “Active and placebo transcranial magnetic stimulation effects on external and internal auditory hallucinations of schizophrenia.” Acta. Psychiatr. Scand., vol. 135(3), pp. 228-238, 2017.

[3] Chervyakov, A.V., A.Y. Chernyavsky, D. O. Sinitsyn, and M. A. Piradov. "Possible mechanisms underlying the therapeutic effects of transcranial magnetic stimulation." Frontiers in Human Neuroscience, vol. 9, pp. 1-14, 2015.

[4] Alzowelei, M., M. Congdon, J. Havas, T. Wang, Y. Sun, and B. Silver. “A lightweight helmet with rotating permanent magnets for brain injury rehabilitation.” 42nd Northeast Bioengineering Conference, Binghamton University, Vestal, NY, April 5-7, 2016.

Radio Frequency Identification for the Development of a Smartphone Integrated Monitoring System for Hand Hygiene Compliance

*Tou Khang, Jacques Dorval, Esteban Tamayo, Ying Sun, Ph.D., Brian Silver, MD**
Department of Electrical, Computer, and Biomedical Engineering, University of Rhode Island, Kingston, RI 02881;
* University of Massachusetts Medical School, Worcester, MA. Correspondence email: yingsun@uri.edu

***Abstract*—In many different medical environments, there is a large risk of healthcare-associated infections (HAI). Thus the project is concerned with the development of a system to monitor the hand-washing performance of healthcare professionals. A monitoring system using radio-frequency identification (RFID), water sensors, an Arduino microprocessor, and communication capacities of a smartphone was developed in this pursuit. An Arduino microprocessor was programmed to access RFID cards within the vicinity of the device scanner. The system used the RFID code to monitor the hand-washing activity that should occur at the designated hygiene location. Inconsistent events would then be sent to management via a smartphone application. A prototype was developed that accomplished the aforementioned goals. The proposed smartphone-based system provides a convenient and cost-effective solution to monitoring hand hygiene compliance.**

Keywords—radio-frequency identification (RFID), healthcare-associated infections, hand hygiene compliance

I. Introduction

Throughout developed nations, there is approximately between 5% and 10% of patients that acquire health care-associated infections, and it is estimated that 15%-40% of those admitted to intensive care suffer from HAI [1]. The World Health Organization also reports that there are approximately 90,000 deaths in the United States, with an estimated cost of $4.5 - 5.7 billion.

A collection of studies have found that when measures are taken to increase compliance with HHC regulations, that the number of those that comply with regulation increase with statistical significance [2]. Out of the studies covered by the previous reference, 58% of the included studies found a significant reduction in HAI when compared with a control group. Another study had observed that before monitoring HHC compliance, that the rates of compliance were 26% for intensive care units and 36% for non-intensive care units [3]. After a year of monitoring and observing, the study found the intensive care units rates of compliance increased to 37%, and non-intensive care units increased to 51%.

There have been several patents that discuss different approaches to implement monitoring methods for the goal of increasing hand-hygiene compliance (HHC). One patent in particular discussed the usage of radio frequency identification (RFID) technology to ensure that individuals with RFID cards enter a room must wash their hands while complying to hand-washing regulations [4]. Utilizing the basis of a RFID tracking system, the aim is to implement a system that tracks the HHC compliance with extra conditions. There were a few systems that appeared to be based on a complex system located with the professionals, not with the rooms. One system, named the nGage™ (Preventix, Hoorn, the Netherlands) utilized RFID tracking and dispenser sensors to monitor compliance [5]. The project is intended to take advantage of the ubiquitous smartphones and their communication capacities, which will allow for a simpler and cost-effective solution for immediate notification of HHC.

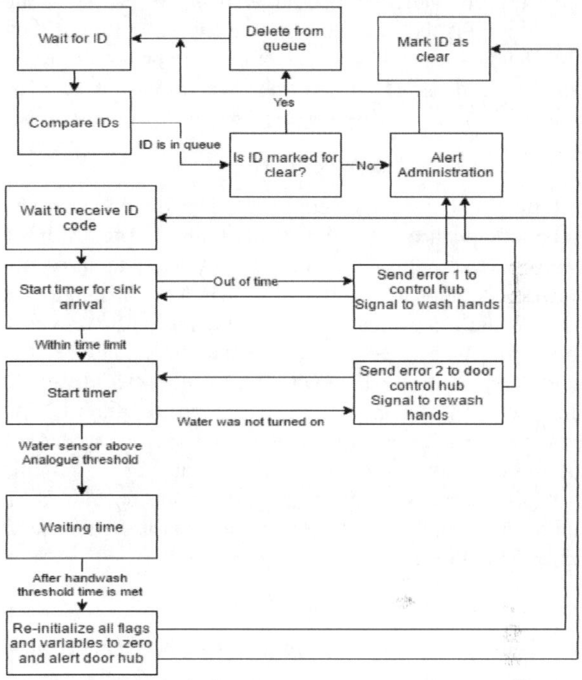

Figure 1. Flowchart of the proposed HHC monitoring system.

II. Methods

A. Tracking

As shown in Fig. 1, the tracking component of the system is the RFID cards that would be carried by healthcare professionals. When an individual enters the room, a hub at the door with a RFID sensor receives the identification code (IDC), then enters the IDC into a queue. Upon scanning the

IDC, the sink module initializes timers. Once an inconsistent event is detected, the system sends the IDC and an error code back to the central hub and then reports the event to management. If the individuals wash their hands following the HHC compliance, the IDC will not be transmitted back to the hub.

B. Evidence

The most basic option is a timer. This timer holds the individuals to HHC standards in the healthcare field, and requires enough time be spent on the hands. The next form of evidence is a sensor that detects the presence of water. The sensor prevents individuals from arriving at the sink and waiting there for the timer to expire while not washing their hands. The sensor is coded to detect whether or not the water faucet has been turned on and track the flow of water. The flowchart in Fig. 1 specifies the event sequence of the system.

III. Results

A system and system structure was developed in order to monitor compliance in a medical setting. The Arduino processors were connected to devices that detect the presence of water, communicate via Bluetooth wireless channels, give indications via color-coded LED's, detect inconsistent events, and transmit the error codes out to a smartphone device. The network tracks hand-washing with timers and water sensors to ensure that hand washing occurs. If an error occurs, the sink system sends the door hub the IDC and the error code. A queue system was created to allow for tracking of multiple individuals. A basic smartphone application was developed for communication and alerts. The layout of the prototype systems are given in Figure 2.

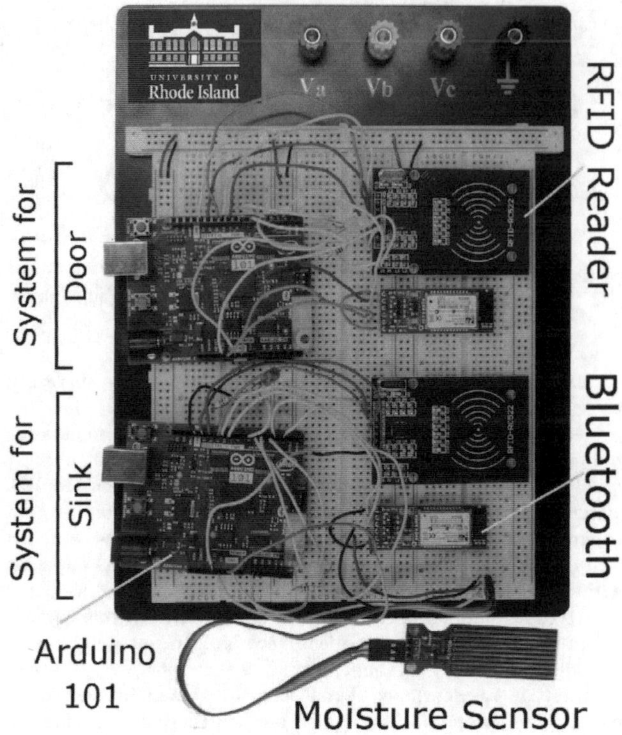

Figure 2. Prototype systems for HHC monitoring.

IV. Discussion

A prototype system for monitoring the hand-hygiene compliance by using the RFID technology and smartphones has been developed. While the system provides the basic functions to achieve the goal, a few conditions were not observed or accounted for in the programming. The queue system allows for members to enter the room while the system is able to monitor whichever individual person washes their hands. When someone enters the room, there is always the chance that they leave the room, realizing that they forgot something or that they potentially entered the wrong room, and thus do not need to wash their hands. The code to correct this situation has not yet been implemented in the software and can be improved as follows: When a duplicated IDC was sent by the door hub and the individual has not appeared at the sink's RFID sensor, a cancellation condition is met and the IDC is removed from the queue. When the individual finally exits the room, the system compares the IDC with the existing queue, then checks to see if the individual has washed their hands. If they have, their IDC's are deleted from the queue. Otherwise, their IDC's are reported to management.

A key development of the system structure is to advance what is already on the market. The main development by this system is the capacity to report HHC violations directly to management on a timely basis. Since the system can report the inconsistent events via a smartphone application, management of the HHC is not restricted by locality and accessibility of the hospital database or network. The instantaneousness and mobility of the system developed in the project allows for responses where the previously developed systems may not be able to achieve.

References

[1] World Health Organization. (2007, May). "Improved hand hygiene to prevent healthcare-associated infections," *Patient Safety Solutions*, vol. 1, Sol. 9, [Online], Available: http://http://www.who.int/patientsafety/solutions/patientsafety/PS-Solution9.pdf.

[2] P. G. Shekelle, *et al.*, "Making health care safer II: an updated critical analysis of the evidence for patient safety practices," RAND Corp., Santa Monica, CA, Evidence Rep. 211, Mar. 2013.

[3] M. McGuckin, R. Waterman, and J. Govednik, "Hand hygiene compliance rates in the United States—a one-year multicenter collaboration using product/volume usage measurement and feedback." *American Journal of Medical Quality*, vol. 24. pp. 205-213, 2009.

[4] R. C. Johnson, "Apparatus and method for monitoring hand washing," U.S. Patent 6 038 331, Mar. 14, 2000.

[5] R. Yarbrough, *et al*, "Efficacy of nGage™ by Proventix, an Electronic Hand Hygiene Surveillance and Feedback Monitoring Device, Against Healthcare Associated Infections," [Online], Available: http://www.proventix.com/clientuploads/library/clients/PBMC-White-Paper-2010-Efficacy-of-nGage-by-Proventix-an-Electronic-Hand-Hygiene-Surveillance-and-Feedback-Monitoring-Device-Against-HAI.

Radio Frequency Identification and Mobile Technologies for Personnel Tracking in a Hospital Environment

*Robert Thottam, Joshua Powers, Michael McAfee, Tanya Wang, Ying Sun, PhD, Brian Silver, MD**
Department of Electrical, Computer and Biomedical Engineering, University of Rhode Island, Kingston, RI 02881, USA;
*University of Massachusetts Medical School, Worcester, MA. Correspondence email: yingsun@uri.edu

Abstract— **The purpose of this design project is to develop a system for tracking patients and staff members in a hospital environment by using the radio frequency identification (RFID) and smartphone technologies. A RFID system includes a reader and numerous tags containing individual identification numbers. These tags can communicate with a mounted tag interrogator using radio waves. The system is intended to be installed at the entrances of specific rooms or areas, allowing for real-time tracking of persons carrying the RFID tags. The focus of this endeavor is on the integration of an Android smartphone platform that provides instantaneous information of hospital personnel, anywhere and anytime. The prototype system of this project will provide a running record of patient/staff locations with tag identifications, to be displayed on an Android application and shared through Bluetooth communications.**

Keywords—radio frequency identification, Bluetooth, hospital environment, patient and staff tracking, Android application

I. Introduction

Topics related to tracking patients and staff in hospital environments have been studied for the past three decades [1]-[3]. Various approaches have been exploited, such as telephone [4], dedicated radio-frequency transmitter [5], motion and pressure sensors [6], radio frequency identification (RFID) [7], and wearable sensors [8]. Among all available technologies, the RFID technology seems to be the most promising [9],[10].

First introduced in 1948, the RFID technology has taken many years to become affordable, available, and reliable for widespread use [11]. Devices that utilize RFID technology fall into one of two broad categories; those with a power supply, known as "active tags," and those without a power supply, or "passive tags." Active tags are almost always more expensive and larger in size. These type of tags contain a battery that can last up to 10 years. Active tags typically operate at a higher frequency, and in turn can be "read" or identified from a longer range. Passive tags are inexpensive, smaller, and often have an unlimited life. However this type of tag has limited data storage capability and a shorter read range [11].

RFID systems operate at different frequencies, ranging from 100 KHz to 900 MHz. Low frequency tags operate from 100-125 KHz, high frequency tags operate at approximately 13.56 MHz, and ultra-high frequency systems operate as high as 850-900 MHz. Low frequency tags are inexpensive, use less power, and are better able to penetrate non-metallic substances. These tags are ideal for scanning objects at close range (less than a foot) [12].

The application of the smartphone technology is still relatively new to healthcare [13]. The mobile health (mHealth) technology involves the use of smartphones in medical care, with applications to areas such as preventive health care services and lifestyle self-management in individuals with chronic illnesses [14].

This project aims at bringing the RFID technology and the smartphone technology together in order to create a platform for personnel tracking in a hospital environment.

II. Methods

A. Hardware

As shown in Fig. 1, the hardware of the prototype system consists of a microprocessor (Arduino UNO), a RFID tag interrogator (MIFARE RFID-RC522, NXP Semiconductors, Eindhoven, Netherlands), two Bluetooth modems (BlueSMiRF Silver, SparkFun, Boulder, CO), all powered by a portable USB charger. An Android tablet is used to communicate with the RFID interrogator and for developing the Android application.

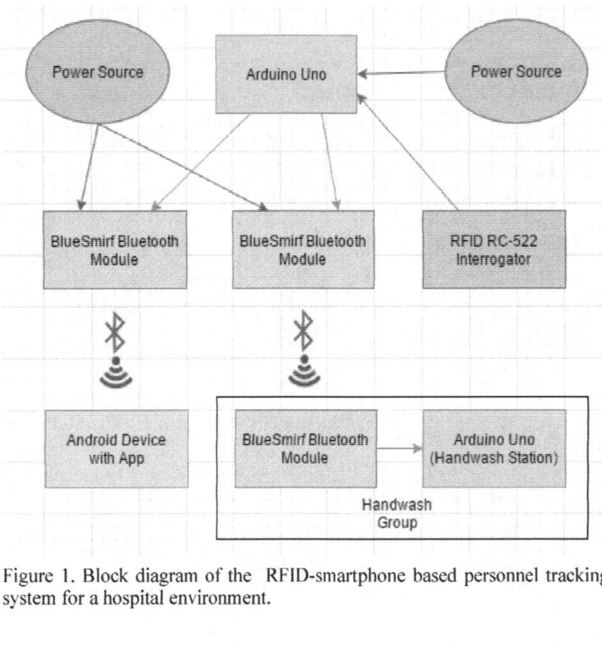

Figure 1. Block diagram of the RFID-smartphone based personnel tracking system for a hospital environment.

Figure 2: The functional prototype of the RFID-smartphone based personnel tracking system. During the development, an Android tablet was used instead of a smartphone.

B. Arduino Software

The software for controlling the RFID interrogator was programmed into the Arduino processor by use of the Arduino Integrated Development Environment (IDE version 1.6.13). The code was written in the C++ language.

C. Android Software / Application

On the smartphone side, the Android operating system was chosen due to its affordability and compatibility with the hardware. The development tool was the Android Studio (version 2.2) and the programming language was Java.

III. RESULTS

Fig. 2 shows the functional prototype of the personnel tracking system, realized on a breadboard and with an Android tablet. A list of the personnel's names and the assigned RFID tag numbers are stored in the Android tablet. When a tag is scanned by the RFID interrogator, the tag ID number is transmitted to the tablet via a Bluetooth link. The tablet looks up the person's name based on the received ID, and displays the name of the person on the screen.

IV. DISCUSSION

A prototype system has been successfully developed to identify personnel based on their RFID tags as they go through a specific doorway. The major advantage of this system pertains to its affordability. The system is also easy and flexible to install because of its capability of wireless communications. These advantages are important considering the need to stall many such units in order to track the personnel throughout a hospital. The RFID interrogators would need to be installed at various room doors and strategic points.

The drawback of the present prototype is the limited range (1-10 centimeters) of the low-frequency RFID reader. To increase the scanning range, an higher frequency RFID reader should be used. The typical range is up to 1 m for the high-frequency RFID reader (13.56 MHz), and about 5-6 meters for the ultra-high-frequency RFID reader (865-960 MHz).

For future work, a Wi-Fi communication link to a hospital server should be established. This would allow the Android app to download its ID database from and upload the location data to the hospital server.

ACKNOWLEDGEMENT

The authors would like to express their gratitude to the University of Rhode Island 2016-2017 Undergraduate Research Initiative Awards Program for recognizing and funding this project.

REFERENCES

[1] Chapman, L. J. et al. "Patient tracking system," Radiological Society of North America 73rd scientific assembly and annual meeting, Nov. 29 – Dec . 4, 1987.

[2] Giles, L., D. Barton, and P. Pope. "Patient tracking system for hospital emergency facility," US 5760704 A, 1998.

[3] Bowen, M. E., C. A. Wingrave, A. Klanchar, and J. Craighead. "Tracking technology: lessons learned in two health care sites," *Technol. Health Care,* vol. 21(3), pp. 191-197, 2013.

[4] Katz, P. "Telephone-based personnel tracking system," US 5255183 A, 1993.

[5] Hawkins, R. E., and M. D. Burke. "Patient tracking system," US 4814751 A, 1989.

[6] Huang, A. Y., G. Joerger, R. Salmon, B. Dunkin, V. Sherman, B. L. Bass, and M. Garbe. "A robust and non-obtrusive automatic event tracking system for operating room management to improve patient care," *Surg. Endosc.,* vol. 30(8), pp. 3638-3645, 2016.

[7] Chen, J. C., and J. J. Collins. "Creation of a RFID based real time tracking (R-RTT) system for small healthcare clinics," *J. Med. Syst.*, vol. 36(6), pp. 3851-3860, 2012.

[8] Gamboa, H, F. Silva, and H. Silva. "Patient tracking system," 4th International Conference on Pervasive Computing Technologies for Healthcare (PervasiveHealth), March 22-25, 2010.

[9] Iadanza, E., F. Dori,, R. Miniati, and R. Bonaiuti. "Patients tracking and identifying inside hospital: a multilayer method to plan an RFId solution," Proc. IEEE Eng. Med. Biol. Soc., pp. 1462-1465, 2008.

[10] Fry, E. A., and L. A. Lenert. "MASCAL: RFID tracking of patients, staff and equipment to enhance hospital response to mass casualty events," Am. Med. Informatics Assoc. (AMIA) Annu. Symp. Proc., pp. 261-265, 2005.

[11] Roberts, C. M. "Radio frequency identification (RFID)," *Computers & Security*, vol. 25, no. 1, pp. 18–26, Feb. 2006.

[12] Zhu, S. K. Mukhopadhyay, and H. Kurata, "A review of RFID technology and its managerial applications in different industries," *Journal of Engineering and Technology Management*, vol. 29, no. 1, pp. 152–167, Jan. 2012.

[13] Dimitrov, D. V. "Medical Internet of things and big data in healthcare," *Healthc. Inform. Res.,* vol. 22(3), 156-63, 2016.

[14] Hartzler, A. L. et al. Acceptability of a team-based mobile health (mHealth) application for lifestyle self-management in individuals with chronic illnesses," Proc. IEEE Eng. Med. Biol. Soc.. Aug. 2016.

The Implementation of Safety Systems into Ride-On Cars to Enhance Play Therapy for Children with Physical Disabilities

Celia Dunn, Alaa Eid, Miranda Mitchell, Joseph Maestri, Ying Sun, Ph.D.
Biomedical Engineering Program, Department of Electrical, Computer, and Biomedical Engineering
University of Rhode Island, Kingston, RI 02881. Correspondence: yingsun@uri.edu

***Abstract* - Children with mobility impairments experience less independence and freedom than other children. To solve this problem, the Lil' Rhody Riders project was founded between the physical therapy and biomedical engineering departments at University of Rhode Island (URI). Ride-on cars were adapted for children with mobility impairments to allow them to experience more independence, freedom, and interact with their peers. The most important aspect of modifying the ride-on cars is safety; cars were equipped with safety straps to aid in proper posture and positioning of limbs. The child's controls of the car were modified to simpler systems to fit each child's needs. To add additional safety, automatic collision avoidance systems using ultrasonic sensors, PIC microprocessors, C++ coding, and relays were implemented onto the ride-on car's electrical system. A parental override Android application was also developed using Android Studio and implemented using Bluetooth and the PIC microprocessor. A prototype was developed with these features and offered a model for a commercialized ride-on car for children with mobility impairments.**

Keywords—mobility impairments; therapy; ultrasonic collision avoidance; Bluetooth; ride-on cars; Android

I. Introduction

Cerebral Palsy (CP) is the most common childhood disability. Around 500,000 children are currently living with CP. Studies have shown that about 3 out of every 1,000 babies born will be diagnosed with CP and roughly 8,000-10,000 babies or infants are diagnosed with CP each year. Similar to other mobility impairments, children with CP experience trouble with muscle coordination and control over motor movements [1]. Children with CP can benefit from various types of physical therapy; helping to overcome their limitations and improve muscle function is among the numerous benefits. One study involving 55 children with CP found that when examining functional skills in daily situations, children exposed to play therapy improved more than in traditional therapy [2]. Additionally, a study by Huang et al. [3] concluded "modified [ride-on] cars have a serious potential to be a fun and functional power mobility option for children with special needs". Commercially available ride-on cars have dual-task control systems that require the gas foot pedal and the steering wheel to be used simultaneously. The lack of muscle coordination in a child with a neurological disorder can make this task seem nearly impossible.

A multidisciplinary team of biomedical engineers and physical therapy students was formed at URI to solve this problem by integrating play and physical therapy. The team adapted ride-on cars for children in the community with CP and other mobility impairments. The prototype for this project included additional posture support with safety straps. The dual-task control system was rewired in to one task. An ultrasonic automatic collision avoidance system was also implemented to compensate for delayed reflexes (Fig. 1.). Emergency parental controls were created in an Android application to allow the parents to remotely stop the car using Bluetooth technology.

The modified ride-on car improved the child's sense of inclusion and independence through play and their mobility through physical therapy exercises. Each individual car was built depending on the child's dimensions, anticipated physical therapy exercises, and parental wishes. As this project progressed, the adaption of ride-on cars continued.

II. Materials and Methods

A. Electrical design

One of the first steps taken to modify the ride-on car was to rewire the original wire control system to meet the specific requirements. A block diagram of the system can be seen below in Fig. 2. By rewiring the leads to the gas pedal, a single button could control the gas. Additional modifications to the car were made so that a joystick could be used instead of a wheel for simplified steering. This was achieved by tapping into the remote-control circuit so that the joystick is dictated by electrical signals rather than simple mechanics. The power to the car was rewired to our own integrated circuit board that utilized the ultrasonic sensors.

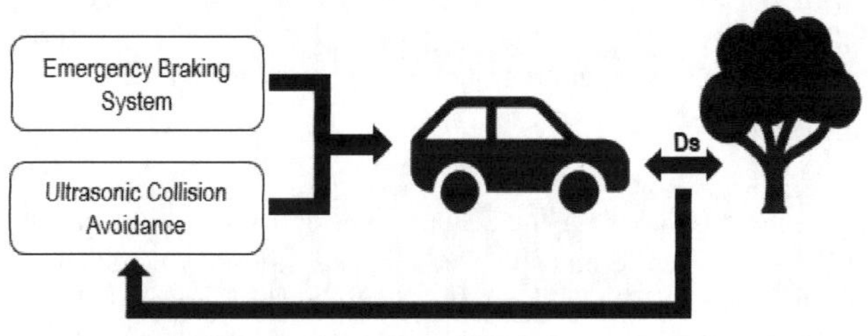

Fig. 1. Diagram of Incorporated Safety Features, "Ds" is Programmable Stopping Distance

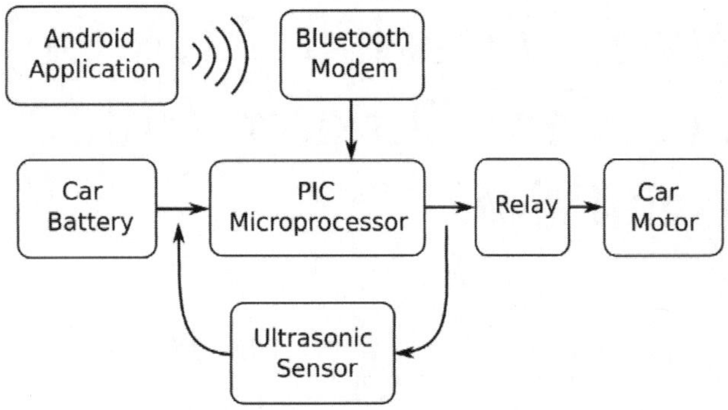

Fig. 2. Block Diagram of Adaptions in Ride-On Car

Using a PIC microprocessor, we can control the ultrasonic sensor and determine how the car reacts. From the output of the PIC, the signal from the electronics is not enough to power the entire vehicle so a relay was implemented allowing the low current electronics to control our high current car. Capacitors help retain the charge, which allows the car to remain powered without restarting.

B. Software Design

To implement the collision avoidance system, a program was developed in MPLABX (Microchip, Chandler, AZ), which programmed the PIC microprocessor to cut the power to the gas when the car was three feet or less away from an object in the distance. The program calculates the distance by computing the length of the echo received by ultrasonic system, and converts this to feet. The ultrasonic sends a signal out every 100 milliseconds, triggering fast enough to keep the child safe, but also slow enough to ensure there is enough time to complete the calculations. Another safety feature added to the car was the development of an Android application used as an emergency brake. An application was created through Android Studio with a single button for stop. Once the button is pressed, this information is relayed to the PIC microprocessor via Bluetooth, and the PIC is programmed to read this information and tell the car what to do.

III. Results

The ultrasonic automatic collision avoidance system was integrated into the car and successfully stopped the car (see Table I). The total cost of this system was approximately $15. Lastly, the Android application successfully communicated with the automatic collision avoidance integrated circuit and stopped the car (see Table I). The total cost of the Bluetooth Android communication system was approximately $25.

IV. Discussion

The main goal for the adapted ride-on cars is to ensure the safety of the children. The needs for the ride-on cars vary with each child. Future directions include the support design adjustments for each child, depending on their needs, dimensions, and personal goals. Throughout this project, we experimented with numerous hardware modifications, we learned how to rewire the gas to a simple pushbutton, and how to rewire the steering system to a joystick. These skills can be applied when changing the hardware in the other cars. By understanding the underlying system of the ride-on cars, rewiring the additional cars and making the adjustments will be rather simple. The software applications will be universal from car to car. However, before the collision avoidance system and emergency brake is implemented on future cars, more tests need to be run to ensure their prolonged safety. In the future, the implementation of a graded joystick with jitter control would be ideal for the car, especially for children with little to no control of muscle coordination [4].

TABLE I. Time and Distance Delay *

	Time Delay (seconds)	Distance Delay (cm)
Ultrasonic Collision Avoidance System	0.51 ± 0.10	46 ± 6.1
Android Application Emergency Stop	0.33 ± 0.00	29 ± 2.8

*The time (seconds) and distance (cm) delay of the ride-on car stopping at 2 mph and Ds was set to 91 cm.

A major technical contribution of this project is the development of the Android application for remote parental controls of the ride-on cars. The experimental data showed that the Android wireless platform provided a good response time to ensure safe operation of the ride-on cars. For this project, the Android application can only stop the car, future projects should enable the application to control direction of the vehicle and start the car as well. The prevalence of the Android smartphones and tablets is advantageous for the implementation of various safety features in a cost-effective way.

Acknowledgment

This project was supported in part by a grant from 2016-2017 Undergraduate Research Initiative Awards Program of the Division of Research & Economic Development, University of Rhode Island.

References

[1] K. Krigger, "Cerebral palsy: an overview," American Family Physician, vol. 73(1), pp. 91-100, 2006.

[2] M. Ketelaar, A. Vermeer, H. Hart, E. Van Petegem-van Beek, P. J. Helders, "Effects of a functional therapy program on motor abilities of children with cerebral palsy," Physical Therapy, vol. 81(9), pp. 1534-1545, 2001.

[3] H. H. Huang, C. B. Ragonesi, T. Stoner, T., Peffley and J. C. Galloway, "Modified toy cars for mobility and socialization: case report of a child with cerebral palsy," Pediatric Physical Therapy, vol. 26(1), pp. 76-84, 2014.

[4] P. Feys, et al., "Assistive technology to improve PC interaction for people with intention tremor," Journal of Rehabilitation Research and Development, vol. 38(2), pp. 235-243, 2001.

Developing an Android Application to Determine Short-Term Induced Heart Rate Variabilities

Leah Acquaviva, Emma Hindinger, Abdullah Albakr, Ying Sun, Ph.D.
Biomedical Engineering Program, Department of Electrical, Computer and Biomedical Engineering
University of Rhode Island, Kingston, RI 02881, USA. Correspondence email: yingsun@uri.edu

Abstract— **Monitoring of the heart rate variability (HRV) can be used to assess the viability of the autonomic nervous system. The portable device described in this paper allows for simplicity in home testing of HRV. By implementing a real-time QRS detection algorithm, the beat-to-beat heart rate (HR) is displayed on either the device's LCD screen or an Android smartphone/tablet via a Bluetooth wireless link. An IRB approved human study was conducted. The study protocol included two simple exercises to induce short-term HRV: the Valsalva maneuver and the sudden standing-up, each during a 15-s period. The preliminary result showed that the degree of HRV varied over a relatively broad range. The beat-to-beat HR had an increase trend during the Valsalva maneuver and an increase-then-decrease trend during the sudden standing-up. The HRV's between Valsalva maneuver and sudden standing-up were similar in range but very weakly correlated (r=0.13, n=10).**

Keywords—heart rate variability; Valsalva maneuver; sudden standing-up; human study; embedded system; Bluetooth; Android application

I. Introduction

Heart rate variability (HRV) is the variation of the beat-to-beat heart rate (HR) or the R-R interval, which is caused by the physiological regulation of the HR via the autonomic nervous system. For healthy subjects, HRV can be influenced by age, gender, body mass index and functional capacity [1]. Many studies have also shown that HRV is further affected by several pathophysiological and environmental factors [2]. HRV is determined over a time period typically between 5 minutes and 24 hours [3]. This study aims at the beat-to-beat trend of HRV during a very short time period, such as 15 s, induced by certain physical exercises.

The issue addressed in this study is the lack of a fast method to assess HRV with the ability to create an understanding of what this number informs us about our health. Fast determination of HRV is important because it can provide a simple indication of the individual's physiological condition. Self-monitoring of HRV has the potential to be a convenient and efficient way to check the health of an individual, especially through the use of a portable personal device. HRV can be established using induced breathing exercises as a method to change the HR like the Valsalva maneuver - attempted forceful exhalation against a closed airway. Previously, a prototype system was developed in our laboratory to facilitate such function [4]. In this project, the portable device was further improved to measure a single channel electrocardiogram (ECG), calculate the HRV, and transmit the data to an android device using a customized application through a Bluetooth wireless link. The users are able to see the trends of their own beat-to-beat HR on their mobile devices.

II. Methods

A. Instrumentation for the HRV Measurement

The block diagram of the system is shown in Fig. 1. The embedded electrocardiogram (ECG) monitor and QRS detector utilizes a customized ECG amplifier and a microprocessor (PIC18F4525, Microchip, Chandler, AZ). The real-time beat-to-beat QRS detection is accomplished by use of the multiplication of backwards differences (MOBD) algorithm [5]. The algorithm was programmed into the microprocessor by using the C language under the Microchip MPLAB X development environment. The program also implements a protocol to measure the beat-to-beat HR during a normal resting state for 15 s followed by an exercised testing state for another 15 s. The normal heart rate (HRn) is the average HR during the normal state. The tested heart rate (HRt) is the maximum HR during the exercised state. The induced HRV is given by: HRV (%) = (HRt – HRn) / HRn x 100%. A Bluetooth modem (blueSMiRF, Sparkfun, Boulder, CO) provides a wireless link at a rate of 115,200 bps to an Android smartphone or tablet. The ECG waveform as well as the beat-to-beat HR are transferred to the Android device for display and further analysis. An Android application was developed by use of the Android Studio integrated development environment (Google, Mountain View, CA).

B. HRV Trend Analysis

The device resulted from this project facilitates the study of exercise-induced HRV and the trend of beat-to-beat HR changes. The exercises used to induce this stress are the Valsalva maneuver and the sudden standing-up motion.

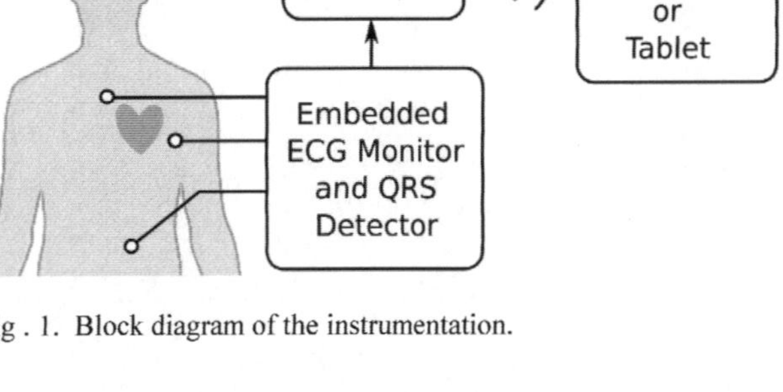

Fig . 1. Block diagram of the instrumentation.

C. *IRB-Approved Human Study*

A human study was conducted with the prior approval from the IRB. In addition to the demographic data, the ECG waveform and the beat-to-beat heart rate were recorded by the microprocessor unit and sent to the Android device for further analysis. The Valsalva maneuver was performed for 15 s. The exhalation pressure was not quantified, nor standardized. The subject was asked to attempt the maximum tolerable force. The sudden standing-up was a movement of standing upright quickly from a sitting position.

III. Results

The embedded instrument was successfully constructed, which incorporated a Bluetooth modem for communicating with an android device. A switch was implemented to direct the data display to either the LCD screen of the embedded instrument or the Android device. The 115,200-bps Bluetooth wireless link provided sufficient bandwidth to transfer the ECG waveform and the HR data to the Android device.

The IRB human research study has been on-going. The preliminary result was obtained from ten (10) anonymous volunteers, as shown in Table I. In the Valsalva maneuver case, HVR = 19% ± 8% (mean ± SD), ranging from 8% to 37%. In the sudden standing-up case, HVR = 20% ± 8%, ranging from 9% to 33%. While the HVR showed similar statistics between Valsalva maneuver and sudden standing-up, the correlation between the two cases was very weak (r = 0.13).

Fig. 2 shows the trends of the beat-to-beat HR changes for Valsalva maneuver (top) and sudden standing-up (bottom). The HR was normalized with respect to the average resting HR. the degree of HRV varied over a relatively broad range. The beat-to-beat HR had an increase trend during the Valsalva maneuver and an increase-then-decrease trend during the sudden standing-up.

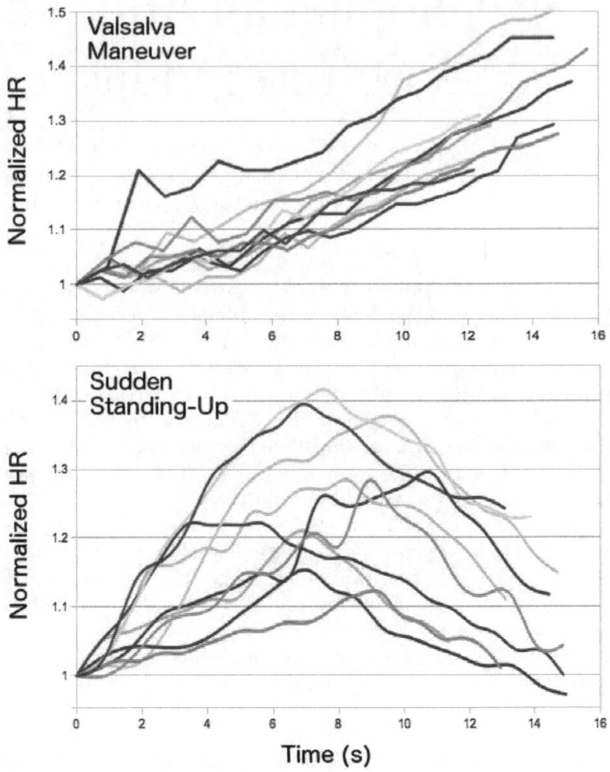

Fig. 2. Normalized beat-to-beat HR for the 10 subjects during the Valsalva maneuver (top) and the sudden standing-up (bottom).

IV. Discussion

The contribution of this project is twofold: 1) developing an Android application for the analysis of the short-term induced heart rate variability, and 2) studying the trends of beat-to-beat heart rate changes during the Valsalva maneuver and the sudden standing-up. The embedded ECG monitor and QRS detector system proves to be an efficient and flexible platform for the front-end instrumentation. The Android based device further provides a powerful tool for waveform display and data analysis. The observed difference in the HR trends seems physiological: The Valsalva maneuver reduces the preload of the heart. Then, the reduced cardiac output activates the baroreflex and increases the heart rate. The suddenly standing-up directly reduces the arterial pressure and thus has a faster activation of the baroreflex.

TABLE I. Demographic and HRV data for the 10 subjects

Subject ID	Gender	Age	HRV (Valsalva)	HRV (Standing)
1	M	25	17 %	17 %
2	F	22	37 %	18 %
3	M	22	11 %	24 %
4	M	22	16 %	9 %
5	M	60	23 %	19 %
6	F	21	20 %	31 %
7	F	21	8 %	17 %
8	M	22	12 %	18 %
9	F	20	23 %	33 %
10	F	22	19 %	11 %

References

[1] Antelmi, I., et al. "Influence of age, gender, body mass index and functional capacity on heart rate variability in a cohert of subjects without heart disease." *American Journal of Cardiology,* vol. 93, pp. 381-385, 2004.

[2] Fatisson, J., V. Oswald, and F. Lalonde. "Influence diagram of physiological and environmental factors affecting heart rate variability: an extended literature overview," *Heart Int.,* vol. 11(1), pp. e32-e40, 2016.

[3] Task Force of The European Society of Cardiology and The North American Society of Pacing and Electrophysiology. "Heart rate variability," *European Heart Journal,* vol. 17, 354–381, 2006.

[4] Junkins, A., C. Thomas, E. Chabot, and Y. Sun. "Heart rate age: baroreflex sensitivity in response to Valsalva maneuver and sudden standing-up," 42nd Northeast Bioengineering Conference, Vestal, NY, April 5-7, 2016.

[5] Sun, Y., S. Suppappola, and T. A. Wrublewski. "Microcontroller-based realtime QRS detection." *Biomedical Instrumentation & Technology,* vol. 26(6), pp. 477-484, 1992.

Impact of Placement of Facial PPG Sensor on Pulse-Rate Monitoring Accuracy

M.K. Bailey, C.J. Smaldone, E. Chabot, Y. Sun
Department of Electrical, Computer and Biomedical Engineering. University of Rhode Island, Kingston, RI 02881

***Abstract*—The facial placement of a PPG sensor can affect pulse-rate monitoring accuracy. In situations where more commonly used sites for pulse monitoring are not available, PPG signals from the face can be taken, and the accuracy of the pulse rate may be affected based upon the PPG sensors facial placement. Using the pulse sensor PPG manufactured by World Famous Electronics LLC., and modifying the raw pulse signal through a series of instrumentation amplifier stages, PPG signals have been both detected and amplified for recording accuracy. Using human clinical trials, this modified pulse sensor will be used to study the impact of different facial placements on the accuracy of the pulse-rate being measured. The results of this research study could prove to aid in the future development of facial PPG sensors as well as their potential applications in wearable devices. The ability to accurately and consistently monitor pulse-rate from different parts of the body such as the face advances the possibilities for numerous applications in the realm of wearable biotechnology.**

I. Introduction

Photoplethysmogram (PPG) sensors are non-invasive electronic sensors used in hospitals, homes, and in many wearable electronic devices to optically obtain a volumetric measurement of an organ. The most common application for using a PPG is with a pulse oximeter, which uses infrared light emitting diodes (LED) to illuminate the skin and measure changes in blood volume through light absorption. The changes detected in blood flow volume are then used to calculate the heart rate or pulse of an individual. In recent years, advances in optical technology have enabled the use of high-intensity green LEDs for PPG sensors. The simplicity, low cost, and reliability of these sensors has led to the increased adoption of this measurement technique [1].

The design concept for the custom facial PPG sensor includes using part of the Pulse Sensor PPG designed by World Famous Electronics LLC. Specifically, the raw pulse output signal obtained through the light reflectance of the high-intensity green LED by the ambient light photodetector is what is needed from the Pulse Sensor PPG [2]. The pulse signal is then amplified and filtered using a custom-built circuit with a series of operational amplifiers.

This modified pulse sensor was used on key locations picked based on the vascular density in relation to human facial anatomy. Testing was conducted in human trials to determine the impact of the placement of facial PPG sensors on pulse-rate.

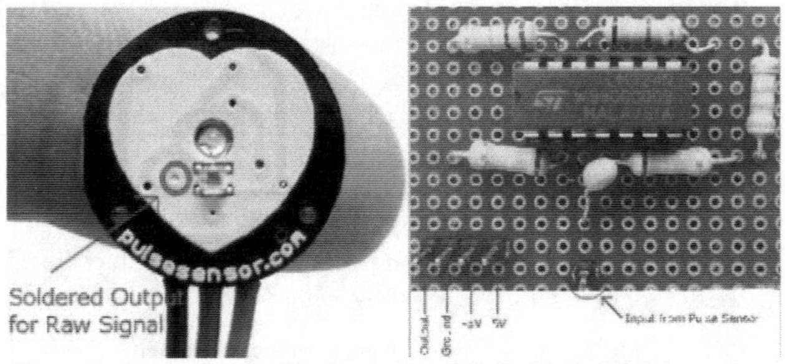

Figure 1: Output Pin (Left), Green Board Gain Stage (Right).

II. Methods

A. Hardware

The PPG sensor being used in this study illuminates the skin using a high-intensity green LED and measures changes in light absorption through the reflected light being bounced back to the ADPS ambient light photo sensor. The light sent through and reflected from the subject measures the blood within the tissue and generates a current in the sensors photodiode. During diastole, the volume of blood in the tissue increases and that amount of change can be measured through light reflectance. The resulting current from reflected light creates a voltage that can be used to calculate the wearer's pulse or heart rate (HR).

The pulse sensor PPG was modified by soldering a wire right at the output from the ADPS ambient light photo sensor to utilize only the raw unfiltered and unamplified pulse signal. However, this signal is weak (has a very low voltage), and so is difficult to be received and analyzed by anything used to process this signal. Therefore, before any filtering or analysis can occur in this research study, the raw pulse signal needs to be amplified.

The amplification of the pulse signal occurs through a series of operational amplifiers (op-amps). Since it is such a small signal (~2mV), it needs to be first AC coupled with a small capacitor in series to be properly read by DC components. The signal is then amplified using multiple stages on a LM324N quad op-amp. The first gain stage increases the signal by 5 times its original amplitude and is soldered onto a green board that is connected directly to the pulse sensor to provide a preliminary "set" gain stage. This amplified signal is then amplified again with another op-amp built on a breadboard that has an adjustable gain between zero and 11 using a potentiometer. The gain is set to increase the signal as much as it can to still be on a 0-5V range. The signal from this amplifier is sent to the PIC microprocessor which is what is used to both record the data during the study and send it to an LED display to provide feedback during recording.

B. Human Clinical Trials

Small scale human clinical trials, supported by an IRB, were used to evaluate the best placement for facial application of the wearable PPG sensor. Based on the vascularity of human facial anatomy, five locations were studied using the sensor. These locations were the left and right ear lobe, the left and right temple, and the forehead. Each location was measured using the sensor attached with a spring-loaded clip or an adjustable strap depending on the location with a contact force of approximately 0.4 N for ideal measurements [3].

The trials studied the effects of the sensor on 20 participants, one participant at a time. Each study took approximately 30 minutes from start to finish. Before the start of the testing, participants were asked to record their height, weight, age, sex, and self-described physical activity level on the questionnaire supplied to them by the researchers. After filling out the questionnaire, participants sat down in the upright position on a chair. The facial PPG sensor was attached using a spring-loaded clip to the thumb to record a baseline PPG signal (for reference). Using three minute intervals, the sensor was placed on the five predetermined locations on the face, one at a time. After each location was successfully measured for the allotted time, the experiment was concluded and the participant was asked to leave the study. The signals were processed using a PIC microchip that is programmed to record the PPG signal, and process it to be shown on an LCD display as a feedback mechanism. Once the data was collected from the clinical trials, the data was transferred to a secure computer where they were furthered analyzed.

III. Results

A. Strength of Signal and Visual Display

The signal was amplified to the ideal voltage with a maximum amplitude between 0 and 5 volts with the ability to adjust some amplification to avoid clipping or overloading the PIC microprocessor using the 2nd gain stage potentiometer. When the signal stayed in the necessary voltage range, the PIC microprocessor successfully recorded the data. From this data, the effect of the facial placements for the PPG sensor for pulse-rate monitoring was successfully studied and analyzed.

The serial data recorded on the PIC microprocessor during the human clinical trials was sent to a file on a secure computer where it was analyzed using MATLAB. All studies were analyzed to determine the most reliable facial placement for accurate and consistent pulse-rates.

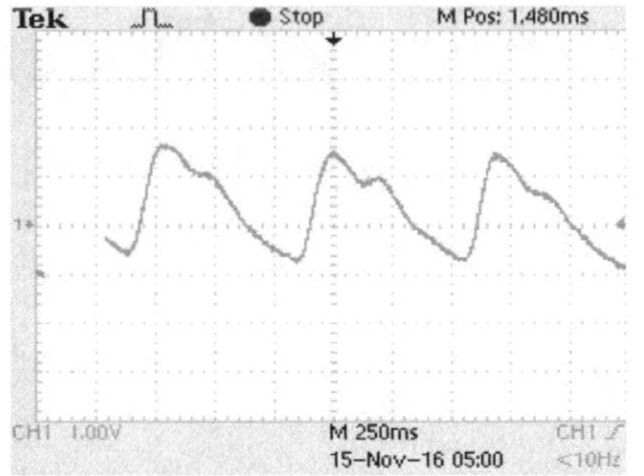

Figure 2: Ideal PPG signal Measured on Oscilloscope.

B. Facial Placement Impact

The expected results from this study are that the earlobes will produce the most consistent and accurate pulse-rate signal as there is already publications with data proving that clean PPG signals are attainable from them. However, it may be noted that this may not be the most ideal spot in all scenarios. To have something pinching and dangling off their ear while they are in constant motion trying to combat situations of distress may be uncomfortable and possibly annoying to the point that they may not wear them. The fact that it is on the peripherals of one face could introduce a lot of noise when even slightly moving one's head, to the point where the signal would be difficult to be read without the proper filtering. There would probably need to be a lot more factors included to provide proper stabilization and accurate readings of the PPG signal in contextual situations as above.

The next best facial placement that would be expected would be the forehead. Although it is not one of the higher vascular density areas of the face, the area between the bone and skin is very thin, and since this study is using reflected PPG measurements, it should still give strong readings because there is less area for the light to be absorbed versus reflected back to the ADPS ambient light photo sensor.

IV. Discussion

A useful platform has been developed for studying technologies using the PPG. The results pertain to the placement of facial PPG sensors should be applicable to wearable devices [4].

References

[1] T. Tamura, Y. Maeda, M. Sekine, and M. Yoshida, "Wearable Photoplethysmographic Sensors—Past and Present.," Electronics, vol. 3, no. 2, pp. 282–302, Apr. 2014.

[2] Murphy, J., & Gitman, Y. (n.d.). Open hardware. Retrieved from http://pulsesensor.com/pages/open-hardware.

[3] X. Teng., Y. Zhang, "Theoretical Study on the Effect of Sensor Contact Force on Pulse Transit Time.," IEEE Transactions on Biomedical Engineering, vol. 54, no. 8, pp. 1490–98, Aug. 2007.

[4] S. Patel, H. Park, P. Bonato, L. Chan, and M. Rodgers, "A Review of Wearable Sensors and Systems with Application in Rehabilitation.," Journal of NeuroEngineering and Rehabilitation, vol. 9, no. 20, Apr. 2012.

Characterizing Ankle Proprioception with Embedded Sensor Balance Board

Thomas Jancura[1], Kelley Magill[1], Ryan Buckley[1], Craig Simpson[2], Eugene Chabot[1]

[1]Department of Electrical, Computer, and Biomedical Engineering

University of Rhode Island, 45 Upper College Road, Kingston RI, 02881

[2]Comprehensive Physical Therapy, North Smithfield, RI 02896

***Abstract*— This paper focuses on the development of a balance board device designed to assist ankle-injury patients during rehabilitation through proprioception training and some initial human study results. The balance board helps in the assessment of the patient's progress by collecting real-time ankle attitude data. Using an Arduino, gyroscope, accelerometer, magnetometer, and a Bluetooth adapter, preliminary results from a human study show similarities in performance between left and right foot proprioceptive tasks.**

I. Introduction

Ankle injuries are one of the most common lower extremity injuries especially in athletes. Many sports require sudden stops and cutting movements with a significant amount of running, such as basketball and soccer. Not only do these ankle injuries result in numerous, high-cost emergency room visits and prevent full participation in sports but they can cause long-term ankle disabilities. Due to this, there is a major financial impact on health care costs and physical therapy. According to the US Consumer Products Safety Commission in 2003, the medical cost of ankle sprains in high school soccer and basketball players was $70 million. [1] Perhaps even more troubling is the long term effects and associated cost. Yeung et al. reports a recurrence rate greater than 70% in athletes. [2]

A common rehabilitation technique post injury involves a balance board for proprioception training. Proprioception training addresses motor control by examining the neuromuscular feedback mechanism that is interrupted with injury, as stated by Lephart. [3] Motor control includes spinal reflexes, cognitive programming, and brainstem activity. Proprioception contributes to the motor programming for neuromuscular control required for precision movements and muscle reflex. After the occurrence of an ankle injury, there is a decrease in sensory input from joint receptors leading to abnormal body positioning, diminished postural reflex responses, and chronic ankle instability leading to the increased probability of re-injury. With proprioception training, dynamic joint stability is obtained and the chance of a reoccurrence of an ankle injury is significantly lowered. [4]

While balance board is pervasive, other methods have been used to quantify the measurement of ankle proprioception. One method, for example, studied the positions of the ankle about three orthogonal anatomical planes as plantar-flexion, dorsiflexion, inversion, eversion, abduction, and adduction. [5] The movements were measured using a 3-Dimensional Goniometry system based on electromagnetic transmitting field. Due to the cost, complexity, and size, the subject must be in a clinical setting, precluding convenient use and monitoring at home or work setting. The device, while shown to have accurate position and pose estimates, requires set up procedures that are not practical outside a rehabilitation setting. Another technique termed Passive Ankle Joint Reposition has been studied ankle proprioception with basketball players by Fu et al [6]. The ankle is placed in various positions such as plantar flexion and dorsiflexion but does not account for the inversion and eversion plane of ankle motions. Many ankle sprains occur in these positions and therefore, this method does not appear to be suitable for full rehabilitation assessment of ankle injuries.

This work describes the development and initial results of a balance board design that is inexpensive and easy to use in a typical clinical or home setting.

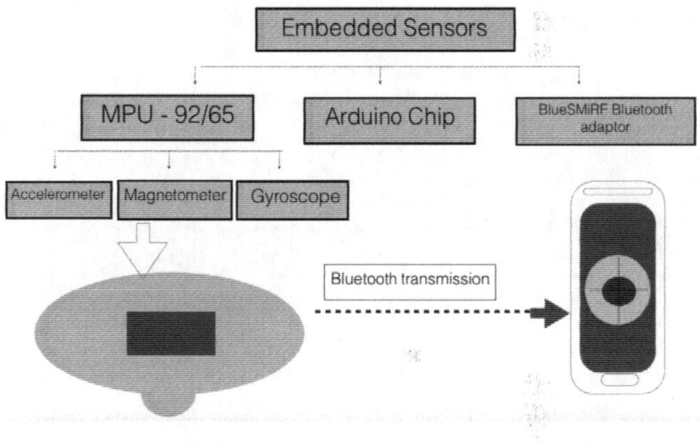

Fig. 1. Block diagram of the balance board embedded with sensor hardware and Bluetooth transmission to the Android application.

IV. Methods

A. Materials and Design Process

The materials used for this project include the Arduino Nano, BlueSMiRF Bluetooth adapter (Sparkfun), MPU9250 (Invensense), a wooden balance board platform with a 3D printed ball (see Fig. 1). The balance board device is paired with an Android device for real-time feedback to the user and data recording. The circuit constructed is placed in the center of the balance board to mitigate the need for real-time transformations of the measurements to a world reference frame. The application displays the orientation of the balance board during training period so the patient can associate the action with the proprioceptive feedback. The application stores all measured data in a comma delimited text format (CSV) for import into data analysis software packages.

B. Clinical Trials

A preliminary clinical trial, approved by the university IRB, was executed to compare ankle positional accuracy of left and right feet using common rehabilitative exercises. The study included individuals who self-reported as having good ankle health. As a benefit of this study, the participants will be more aware of physical therapists' techniques and mechanisms of treatment for strength and proprioception of the ankle. These healthy volunteers will perform multiple movements on the balance board in a seated position and the data will be recorded based on their speed, accuracy, and repeatability of these movements. The subject will place their foot on the board in a seated position with their toes at 12 o'clock and their heel at 6 in the sitting position with hip, knee, and ankle flexed to 90 degrees. The subject is directed by the tester to look down at the round board and press down on the board so that the 12 o'clock position is touching the floor followed by the 6 o'clock position hitting the floor, using their toes and heel. The subject is directed to do this 2 consecutive times and then to stop in the neutral position. Next the tester asks the subject to replicate this motion while looking straight ahead 10 consecutive times. At this point, the data is stored to be compared to the subjects' initial motion while looking down at the board and analyzed by the tester. Next the tester replicates the prior training and testing with side motion (ankle inversion and eversion). Real-time attitude data is collected through the whole training and testing process.

III. Results

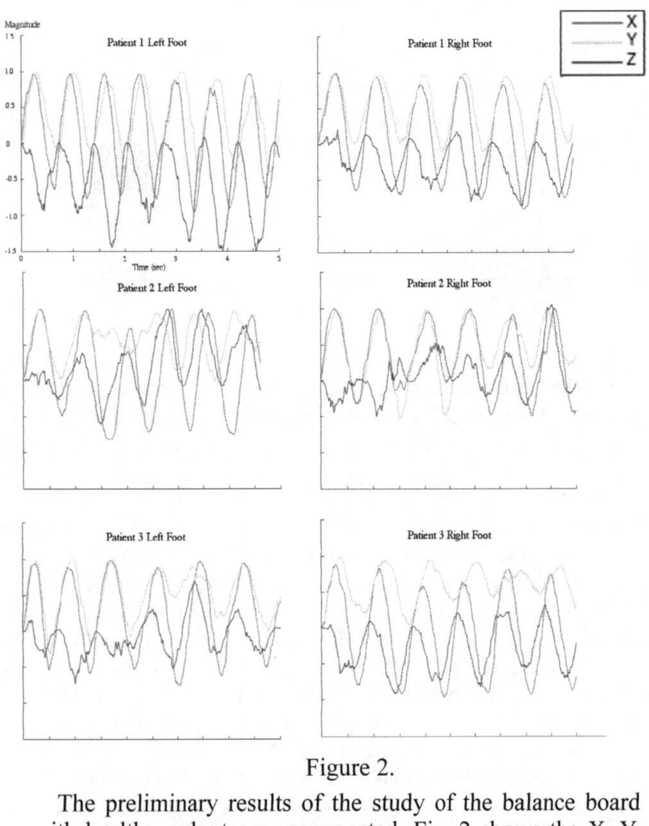

Figure 2.

The preliminary results of the study of the balance board with healthy volunteers are reported. Fig. 2 shows the X, Y, and Z coordinates of the back-and-forth motion of three participants. The right and left columns represent the patients' right and left feet, respectively.

IV. Discussion

Based on the preliminary tests done on the left and right feet of the test subjects, we can conclude that there is a trend showing more control over the right foot, for right hand dominant people. While looking at the graphs, the proper way to compare the two feet is by looking for the similarities in the X and Y coordinates, while maintaining consistency in the Z coordinates of the board. With more consistency in the Z coordinates, this shows the test subject has better control of ankle movement. While Subject 1 shows some consistency in the Z coordinates, the right foot has a greater range of motion while remaining consistent in the z-plane. Subject 2 is less smooth in moving their right foot back and forth than the left. However, like Subject 1, the coordinates of the right foot were more consistent when compared to their left foot. The third subject's right foot shows a better consistency in the Z coordinates, not only in comparison to each other but in the range of motion as well. With IRB approval for human study, further information can be gathered solidify our initial hypothesis.

The balance board is to provide an inexpensive platform for both in office and at home rehabilitation. Qualitative results from the preliminary results shown appear to allude to a correlation between handedness and performance of the corresponding foot. While using acceleration directly may not report an accurate pitch angle of the wobble board, the direction of the pitch appears to be consistent with motion exhibited. The real-time data collected from the study helps to qualitatively assess similarities between left and right feet performance, but further quantitative results are anticipated to characterize the differences in the accuracy, speed, and repeatability of the participants. Understanding normal ankle proprioception is important to diagnosing injured patients, developing procedures to quickly rehabilitate, and potentially reducing associated medical costs. This study overall will not replace a physical therapist but rather will collect knowledge on how to optimize the rehabilitation techniques. Extending prior balance board work [7], a Bluetooth interface is incorporated and self-contained power for convenience of the test administrator or physical therapist. Future work is planned to not only analyze the human study results but to improve the design for a broader field use by physical therapists and patients. In particular, the user interface needs to be improved to allow self-directed data collection, and analyze and present the recorded data for assessment by clinician.

Acknowledgements

This work has been possible thanks to financial support provided by the University of Rhode Island Undergraduate Research Initiative.

References

[1] McGuine, T.A. & Keene, J. S. (2006). The effect of a balance training program on the risk of ankle sprains in high school athletes. *The American journal of sports medicine*, 34(7), 1103-1111.

[2] M. S. Yeung, K. M. Chan, C. H. So, and W. Y. Yuan, "An epidemiological survey on ankle sprain.," Br. J. Sports Med., vol. 28, no. 2, pp. 112–116, Jun. 1994.

[3] Lephart, S. M., Pincivero, D. M., Giraido, J. L., & Fu, F. H (1997). The role of proprioception in the management and rehabilitation of athletic injuries. *The American journal of sports medicine*, 25(1), 130-137.

[4] Mattacola, C. G., & Dwyer, M. K. (2002). Rehabilitation of the Ankle After Acute Sprain or Chronic Instability. *Journal of Athletic Training*, *37*(4), 413–429.

[5] Abboud, R. J., Agarwal, S. K., Rendall, G. C., & Rowley, D. I. (1999). A direct method for quantitative measurement of ankle proprioception. *The foot*, 9(1), 27-30.

[6] Fu, A. S., & Hui-Chan, C. W. (2005). Ankle joint proprioception and postural control in basketball players with bilateral ankle sprains. *The American journal of sports medicine*, *33*(8), 1174-1182.

[7] B. Kotowski, M. Barbin, C. Gomes, C. Simpson, & E. Chabot, "Balance Board with Embedded Sensors for Rehabilitation of Ankle Injuries," in Northeast Bioengineering Conference (NEBEC), 2016 42nd Annual, 2016, pp. 1-2

Pattern Recognition of Dorsal Mounted Linear Vibrotactile Array

Samuel Karnes, John Donahoe, Chris Morino, Eugene Chabot, PhD, and Ying Sun, PhD
Department of Electrical, Computer, and Biomedical Engineering
University of Rhode Island, Kingston, RI 02881-0805 USA

***Abstract*— This paper reports on the findings of a human study assessing the pattern recognition performance when using a dorsal mounted, linear vibrotactile array. These tactile patterns are correlated to three variables: size, speed, and direction. The subjects were trained to identify these vibrotactile responses and then subsequently tested with a series of random sequences. Results from the size, direction and speed experiments reflect that subjects were able to accurately detect the respective variable through the array in 61.5%, 90.5% and 92% of the tests.**

I. INTRODUCTION

According to the American Census survey conducted in 2015, an estimated 7.3 million people in the United States are diagnosed with visual disability [1]. The most common causes of vision loss are due to macular degeneration, glaucoma, detached retina, and diabetic retinopathy [2]. Due to such a large population of individuals affected by such diseases, it is important to have some visual supplements. Most people with severe cases of the aforementioned diseases use the white cane and auditory feedback to get a sense of their surroundings. There are also audio devices that provide information about an individual's surroundings. These devices, however, could still lead to an individual being unaware of the size or speed of an approaching object. Chabot et al. [3] had explored the use of tactile patterns to relay such parameters on the ventral side of the abdomen. This work assesses the pattern recognition performance of subjects using a lumbar mounted linear array varying size, speed, and direction of an active stimulator group. Based upon prior work on the abdomen by Cholewiak [4], it is hypothesized that a dorsal mounted array should have a similar performance to a ventral mounted array. The results, if positive, would emphasize an alternative, potentially less intrusive location for some individuals with tactile communication devices.

II. METHODS

A. *Subjects*

Ten subjects were recruited for the IRB sanctioned human study. Participants were solicited from the university engineering department by email and were not compensated for their time.

B. *Vibrotactile Belt*

A support belt composed of nylon was used in this experiment. Fifteen pockets were sewn on the inner lining to hold each vibrotactile stimulator in place. The stimulators were placed approximately 2.5 cm center to center, which was effective distance for similar patterns as demonstrated by Chabot et al. [3]. All vibrotactile stimulators (Yuesui, B1034.FL45-00- 015), were controlled by a PIC microprocessor (Microchip, PIC18F4520). The training and test array tactile sequences were programmed into the microcontroller.

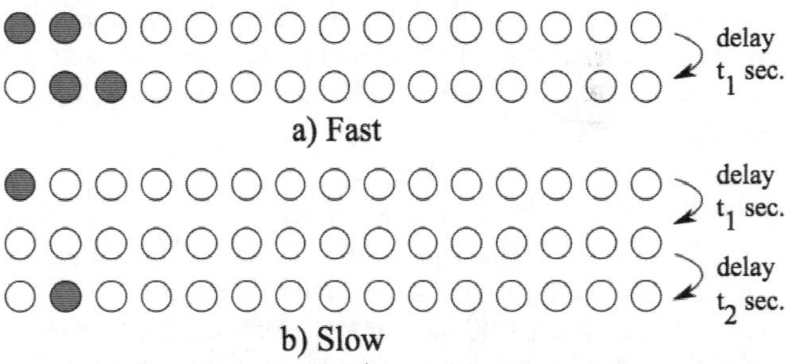

Fig. 1 - Tactile sequences are shown for fast speed (a) and slow speed (b) reflecting two timing parameters: t1 and t2. **Circles represent individual vibrotactile elements. Shaded circles illustrate active elements.** Time t1 denotes the time a stimulator is active, while t2 denotes off-time between patterns

C. *Test Procedure*

The human study was decomposed into three sections: size, direction, and speed testing. Each of the sections was preceded by a training period. Each test involved twenty sequences with a four-second pause after each sequence to allow for responses. The same randomized test sequence was presented to all participants in the same order. For the size variable, every individual was given a small, medium, and then large active stimulator group. Training each subject involved activating these amounts of motors across the right, left, and then middle portion of the belt. Following this training, each subject received the same twenty sets of stimuli with varied location and size. For each of these sequences the individuals were asked by the investigator to report the group size observed. Upon pressing a push button on the electronics, the next training sequence started to present sequences illustrating left and right moving groups. Motors were sequenced on and then back off from right to left (denoted as left moving), and left to right (denoted right moving). The two sequences were repeated once to reinforce the connection between the pattern and desired response. Subjects were asked to provide a response of left or right moving. The final training period focused on the speed parameter. Two speeds were presented: slow and fast moving. Similarly, the responses requested for the test were slow or fast.

A push button was provided to the test conductor to trigger the execution of the next training or test phase. In between segments, the push button was required to progress.

D. *Tactile Patterns*

For the three tests performed, different tactile sequences are presented. The size test reflects a change in the active stimulator group size denoted by small, medium, and large, which corresponds to 1, 2, or 4 stimulators, respectively. The direction test varied moving the active group left or right. The rate of change of this pattern corresponds to the slow sequence in Fig. 1, which defined t1 and t2 as 1 second. Each direction sequence spanned 4 sequential motors. The speed test created a slow or fast moving group (see Fig. 1). The slow delay had each motor turn on for one second and then off for one second before the next started. The fast moving sequence had each motor stay on for two seconds while overlapping the next motor in the sequence.

III. RESULTS

The same ten participants were used for all three tests. The results of each test are presented in confusion matrices Tables 1, 2, and 3 for the size, direction, and speed test, respectively. Additionally, the accuracy of each individual subject is shown in Fig. 2. The mean scores for size, direction, and speed tests are 61.5%, 90.5%, and 92%, respectively. In total, 8 responses were left blank with 6 occurring in the size test and 2 occurring in the direction test.

TABLE 1
Size Test Confusion Matrix

		Responses			Correct %
		Small	Medium	Large	
Answers	Small	48	15	4	68.5
	Medium	12	28	18	46.6
	Large	3	20	47	67.1

TABLE 2
Direction Test Confusion Matrix

		Responses		Correct %
		Left	Right	
Answers	Left	90	10	90
	Right	9	91	91

TABLE 3
Speed Test Confusion Matrix

		Responses		Correct %
		Slow	Fast	
Answers	Slow	95	5	95
	Fast	11	89	89

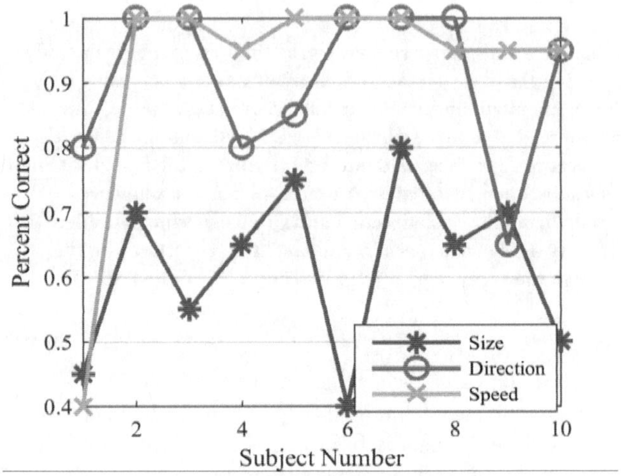

Fig. 2 - Subject Accuracy is shown over all three tests.

IV. DISCUSSION

Vibrotactile arrays provide an inexpensive and low discomfort for use as a tactile communication device. This study reflects a high accuracy in determining patterns, which would be useful in conveying size, speed, and direction. While the size reported by the users was often wrong, the subject was biased toward the correct answer. For example, when the size was small, the subject rarely reported a large size. When used in a real-time feedback device, such as for visual sensory substitution, knowing when objects are approaching or growing in size could be more important than properly judging the absolute size. It is also important to note that the first subject had results indicative of an outlier due to material obstructing the vibrotactile array. The recalculated mean scores, excluding those results, become 63.3%, 92.7%, and 97.8% for size, direction, and speed respectively.

REFERENCES

[1] United States Census Bureau, "2015 American Community Survey 1-Year Estimates."

[2] W. Erickson and S. von Schrader, "2013 Disability Status Report: United States," Cornell University Employment and Disability Institute(EDI), Ithaca, NY, 2014.

[3] E. J. Chabot, "Methods and analysis of a wearable, linear, vibrotactile array to relay motion information for visual sensory substitution," University of Rhode Island, 2011.

[4] R. W. Cholewiak, J. C. Brill, and A. Schwab, "Vibrotactile localization on the abdomen: Effects of place and space," Percept. Psychophys., vol. 66, no. 6, p. 970, 2004.

Integrated EOG and EMG Front-End for Differentiating Intentional and Unintentional Blinks

Mark Plugovoy, Matthew Forde, Tanya Wang, Eugene Chabot and Ying Sun
Department of Electrical, Computer and Biomedical Engineering
University of Rhode Island, 45 Upper College Road Kingston, RI 02881
markplugovoy@my.uri.edu, matthew_forde@my.uri.edu

***Abstract*—Electrooculography (EOG) is beneficial in many human interface systems when the signals are measured, amplified and filtered to produce a signal from movement of eyes by which appliances can be controlled. Some patients who are disabled by paralysis can only control their eyes; therefore, these EOG and electromyography (EMG) signals become one of the few ways they can interact with the surrounding environment. The signals gathered from blinking intentionally and unintentionally, need to be distinguished in order for the system to work correctly. By integrating duration and amplitude data from both EOG and EMG signals simultaneously from three electrodes, an efficient algorithm is proposed to distinguish the two classes of blinking: intentional and unintentional.**

I. Introduction

The main objective of this project is to create an electrical system that can differentiate when the user intentionally blinks versus when he or she unintentionally blinks with the use of electromyography (EMG) and electrooculography (EOG). An electromyogram is a record of the electrical potential from muscle cells. With electrodes attached above and below the eye, an electrical potential is recorded from the orbicularis oculi and levator palpebrae when the patient blinks. An electrooculogram is a record of the electrical potential dipole created by the difference in the electrical potential of the cornea in relation to the retina of the eye. Electrodes are placed around the eye to observe this dipole. Typically, the absolute position of the eye is not measured due to the number of noise sources dominating the potential measured. EMG signal amplitudes range from approximately 0.001 to 100 millivolts with a frequency range of 50 to 3000 Hertz. EOG signal amplitudes range from approximately 0.001 to 0.3 millivolts with a frequency range of 0.1 to 10 Hertz [5]. These ranges are the foundation of our design for filtering and amplifying the signals separately for analysis and use.

Previous work has explored the characteristics of intentional and unintentional blinks. Research by Kaneko [1] demonstrated an ability to isolate the two blink classes using independent electrodes for EMG and EOG signal collection. Our proposed configuration uses three electrodes to record both signals simultaneously. With this configuration, preliminary results appear positive to discern blinking classes with a simple threshold based upon threshold and duration of the events.

II. Build

Fig. 1 shows our schematic for the EMG and EOG bandpass filters along with their respective gain stages and

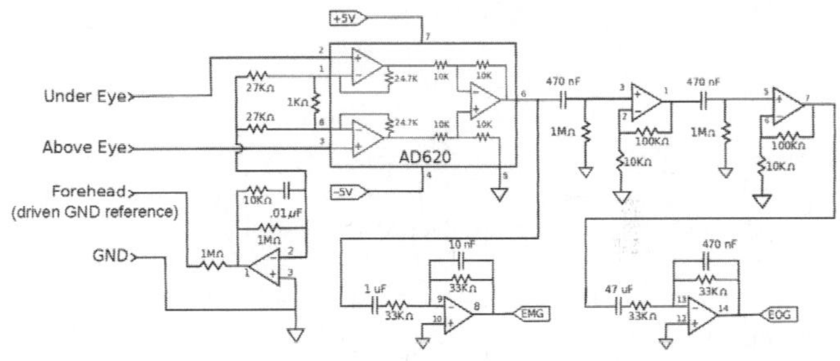

Figure 1: EMG and EOG Circuit.

an instrumentational amplifier (AD620). The bandpass filter for the EMG signal amplifier in our circuit has cutoff frequencies at 50 Hz and 4.8 kHz, putting us in the frequency range of EMG signals. The upper limit of 4.8 kHz was chosen to align with commonly available synthetic capacitors and resistors while also being the closest frequency to the top of the EMG frequency range.

For the EOG bandpass filter, the resistor and capacitor values were chosen for -3 decibel cutoff frequencies at 0.1 Hz and 10 Hz.

For the operational amplifiers, multiple LM324 integrated circuits (ICs) were utilized to provide a driven ground to the forehead, gain stages, and bandpass filters. The LM7660 voltage inverter provided a negative voltage supply for LM324 bipolar operation.

Due to the small signals associated with the EOG, three amplifier stages were implemented to maximize the dynamic range of the signal for measurement on a 0 to 5V input range. Two non-inverting op-amp gain stages are cascaded with the AD620 gain of 50. A total gain of 6,050 was provided with this configuration.

For testing, the AD620 was placed on a small circuit board mounted to the head. The wires were kept short (approximately 8 inches) to reduce the impact of electromagnetic interference.

For EOG and EMG signals, one electrode was placed

above and below the eye. The third electrode on the forehead provided a driven ground reference. This arrangement of electrodes is similar to configurations commonly reported by researchers, such as Kaneko [1], when trying to detect EOG signals. When someone looks up or down, the electrodes will propagate electrical potential changes related to dipole pose modifications. Vertical movements were the focus of the measurements due to the greater amplitudes observed compared to the lateral motion for horizontally placed electrodes.

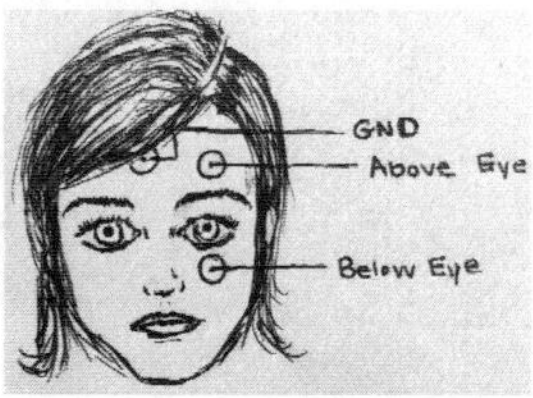

Figure 2: Electrode Placement.

III. RESULTS

After testing each gain stage sequentially, we did a frequency sweep to find where the half power, or -3db point occurs with the hardware setup. We found that the EMG circuit has cutoff points at approximately 70 Hz and 4.3 kHz. We found that the EOG circuit has cutoff points at approximately 2 Hz and 11 Hz.

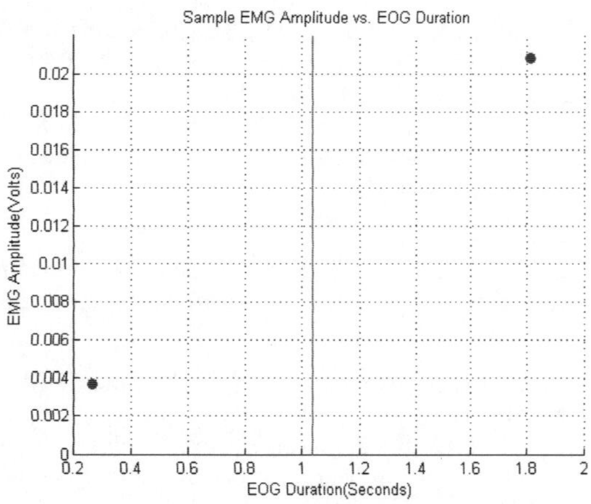

Figure 3: Sample Data Average of EMG Amplitude VS. EOG Duration differentiating Intentional from Unintentional blinks

Based upon example waveforms, a threshold was developed that is applied to the duration of the EOG and amplitude of the EMG. An initial qualitative assessment was obtained based upon an independent threshold in both the EOG and EMG components.

IV. CONCLUSION

The EOG and EMG signals play a crucial role in how people with paralysis can communicate. Differentiating between voluntary and involuntary blinking can make the difference for a patient to communicate effectively and accurately. An apparatus was discussed that qualitatively appears to produce results consistent with reports by Kaneko [1], which allude to a feasibility of pursuing a simple threshold based classifier of intentional and unintentional blinking. While our initial proposal is for a simple logical AND operation between a threshold on EOG and EMG, a second option being considered is a threshold based upon the linear combination of EOG and EMG to better isolate the classes. Future work is anticipated to quantify the performance of the proposed approach with a human study.

Theoretically, a bandwidth for our EMG and EOG circuits was designed for 50 and 3000 Hz, and 0.1 and 10 Hz, respectively, but component availability and tolerance of the capacitors and resistors generated a slightly different set of bandwidths. Even though the realized bandwidth was more limited than initially planned, the design appeared to have similar characteristics to those reported by Kaneko [1] and was from a preliminary qualitative assessment effective at discrimination so the simplistic design was maintained. The use of the same electrodes for EMG versus EOG has led to a contamination of the signal by EOG related potentials, but for the purpose proposed, it does not appear to be significant enough to impact results. If so, a higher order filter could be applied to the EMG signal to reduce further. Future work will explore the impact of different filter approaches by applying post-processed higher order digital filters.

REFERENCES

[1] Kaneko, K. & Sakamoto, K. (1999). Evaluation of three types of blinks with the use of electro-oculogram and electromyogram. Perceptual and Motor Skills, 88, 1037-1052

[2] Arthi S.V. & Norman S.R. (2015). Analysis of electrooculography signals for the interface and control of appliances" International Journal of Multidisciplinary and Current Research(IJMCR), Vol.3, pp.87-90.

[3] Prasannaraj, A. & Dharmalingam, R. (2014). Integrated EOG Based Interface to Control Wireless Robot. International Journal of Advanced Information and Communication Technology(IJAICT), Vol.1, Issue 2, pp.241-244.

[4] Nakanishi, M., Mitsukura, Y., Wang, Yijun, Wang, Yu-Te, & Jung, T-P. (2012). Online Voluntary Eye Blink Detection using Electrooculugram. Proceedings of International Symposium on Nonlinear Theory and its Application,pp.114-117.

[5] "Biomedical Engineering Education Portal." National Instruments. N.p.,29Mar.2016.Web.24Sept.2016

Android Application to Prevent Foot Ulcers and Monitor Weight of Diabetic Patients

Brian McHugh, Michelle Bierman, Ryan Brown, Ying Sun, Ph.D., Jiang Wu, Ph.D.
Department of Computer, Electrical and Biomedical Engineering, University of Rhode Island,
Kingston, RI 02881, USA. Correspondence email: yingsun@uri.edu

***Abstract* – This device combines a scale, camera and an Android application to monitor weight as well as the soles of feet to prevent foot ulcers in diabetic patients. Once stepped on, the scale will send the person's weight to a Android app and the app will record this number in order to keep track of progress over time. Then, the patient will take a picture of their right and left foot, respectively, by positioning their feet over a microcontroller with a camera attachment. The photos will then be sent to and stored on the app so the patient can monitor the soles of their feet.**

I. Introduction

The purpose of the device proposed in this paer is to monitor weight and prevent foot ulcers for diabetic patients that suffer from peripheral neuropathy. Diabetic peripheral neuropathy is the most common complication associated with diabetes [1]. Peripheral neuropathy is associated with nerve damage in the hands, arms, legs and/or feet. One of the most dangerous side effects of peripheral neuropathy in the feet is numbness, which can induce the development of ulcers on the soles of patient's feet. These ulcers progress due to the fact that the patient is unaware that there is an open sore [2]. An untreated ulcer can result in infections in the skin or bone, which may lead to amputation of the foot or leg. Furthermore, almost 90% of people living with type 2 diabetes are overweight or are obese [3]. Obesity can promote the development of ulcers due to the amount of pressure on the patient's feet. If a patient develops these ulcers, weight loss may aid in accelerating the healing time and if caught early enough, infection can be avoided. By monitoring weight and the soles of the feet at the same time, a patient can avoid these ulcers as well as benefit from losing weight.

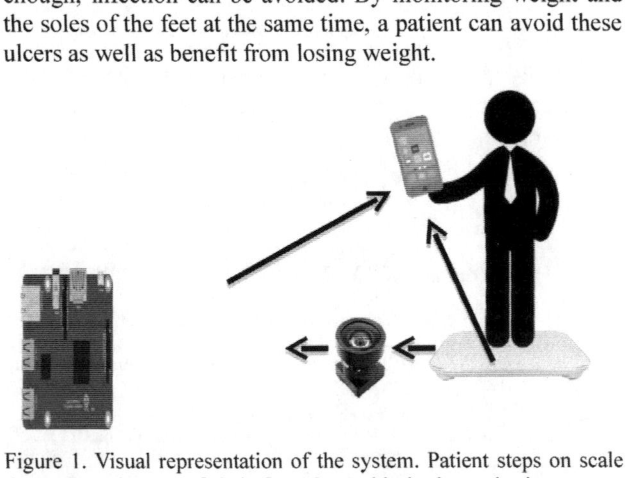

Figure 1. Visual representation of the system. Patient steps on scale then takes pictures of their feet. Once this is done, the images are processed through the Raspberry Pi 3 and then saved on the Android application.

This project is the first multifunctional medical device to preemptively monitor diabetic patient's feet for ulcers. Figure 1 describes the overall block diagram of the device. The combination of a scale, camera, mobile application and microcontroller allows for a patient to have a "one stop shop" product that they can integrate into their daily routines. The goal of this project is to aid diabetic patients who are suffering from peripheral neuropathy in the prevention of a foot ulcer and promote weight loss to benefit their overall health.

II. Methods

A. Hardware

This project incorporates three main hardware components: a camera, a Bluetooth scale and a Raspberry Pi 3 miniature single-board computer. The camera is a wide-angle fish eye lens that attaches directly to the Raspberry Pi 3 Camera Interface port. This allows for the device to take clear photos of the patient's foot that will be used to monitor the soles of the feet. The Bluetooth scale is a Xiaomi Smart Scale XMTZC01HM that meastures a user's weight in pounds and communicates via the Android application. This weight is then stored for later use to track weight loss or gain. The Raspberry Pi 3 with the camera was used to take photos and send images to the Android app. The Raspberry Pi 3 uses a Broadcom BCM2837 chip, powerful enough to process all the necessary information. Together, these components create an all-in-one device to prevent foot ulcers and to monitor weight.

A separate START push button is added to trigger the consecutive execution of these three scripts. Figure 3 shows the connection of this button to the general-purpose input/output (GPIO) connector of the Raspberry Pi 3.

B. Raspberry Pi 3 Software(A, B can be combined)

Two pieces of software were developed, one running on Raspberry Pi 3 and one on Android. The Raspberry Pi 3

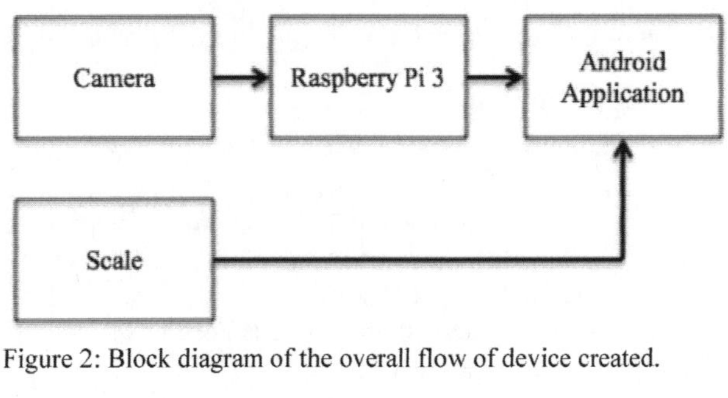

Figure 2: Block diagram of the overall flow of device created.

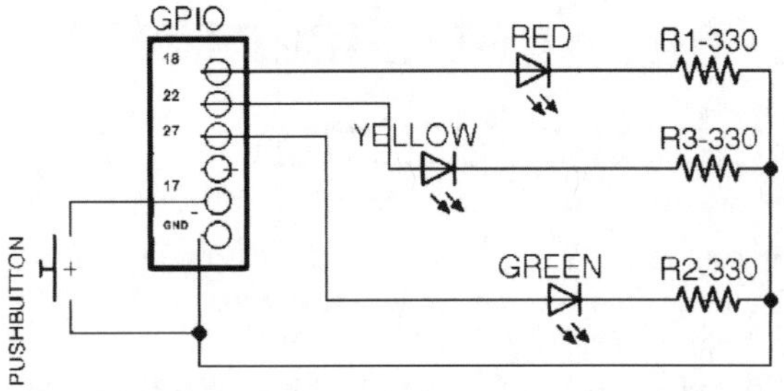

Figure 3: Schematic of circuit used to automate the Raspberry Pi 3.

software was coded using Python, and includes three scripts. The camera.py script takes pictures of the foot. The imagecompression.py script compresses the image that was taken. Finally, an additional script server_sendfile.py creates a network Transmission Control Protocol (TCP) socket to send images to the Android Application.

Once a user presses the START button, the Python program residing in the Raspberry Pi 3 runs, takes a picture of the foot and sends the picture to the Android application. The three different color LEDs light up one after another to indicated the Raspberry Pi 3's current status. The red LED signifies an idle state, the yellow LED denotes the Raspberry Pi 3 is taking a picture and then compressing the image, and the green LED signifies a successful image transfer to the Android application.

C. *Android Software*

The Android application was developed using Android Studio (version 2.2.2). This user-friendly application allows the patient to store information such as photos and weights in one place. The Android software obtains the weight information ofthe user via Bluetooth. A TCP network protocol is used for the image transfer from Raspberry Pi 3 to the Android application through a local Wi-Fi access point.

III. RESULTS

Figure 4 shows the Android application displaying the weight of the patient when stepping on the Xiaomi scale. This weight unit can be chosen between pounds or kilograms, depending on what the patient prefers.

Figure 4 also shows where foot image is displayed. The photo taken by the user will replace the image of the two animated feet (it is better to have foot image). This image is then stored for review later. By clicking the "photos" button, the user will be taken to a database where he or she can view images and weights from previous days.

In the future, patients will integrate this device into his or her daily routines so that these saved images can be compared. Once the images are compared, the patient will be able to see subtle changes to the soles of their feet. This will allow the start of a sore to be oticed before a foot ulcer develops. Patients can then decide if further actions need to be taken in order to prevent the development of a foot ulcer.

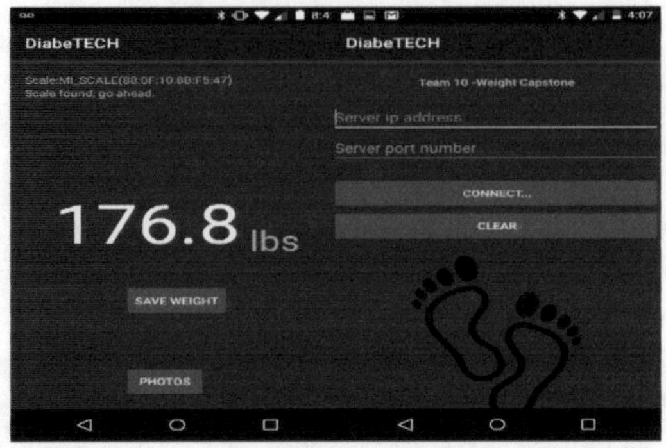

Figure 4: Screenshots of the Android application.

IV. DISCUSSION

This project focuses on a major health issue in today's society. Diabetic foot ulcers are the number one non-traumatic cause for lower limb amputation in the US [4]. This not only lessens a patient's quality of life but also creates a burden on the healthcare industry due to the fact that it costs up to $45,000 dollars each time a patient is hospitalized for diabetic foot ulcers [4].

There are no other preventative foot ulcer measures similar to this device currently on the market. Patients are able to easily compare the soles of their feet over time. A further improvement will be an algorithm that is able to highlight any changes in the surface of the foot and notify the user. If the patient notices any differences in the pictures, they are able to preemptively seek medical attention before an ulcer develops, thus avoiding infection and hospitalization. Many foot ulcers start to develop due to irritation and go unnoticed because of nerve damage. The solution is usually simple, as easy as purchasing new shoes.

Diabetes is a growing problem in today's society and this project is working to use modern technology to help these patients avoid hospitalization and ultimately retain their quality of life.

ACKNOWLEDGMENT

Thanks are paid to the University of Rhode Island Undergraduate Research Initiative program for recognizing and funding this project.

REFERENCES

[1] K. Juster-Switlyk and A. Smith, "Updates in diabetic peripheral neuropathy", *F1000Research*, vol. 5, p. 738, 2016.
[2] W. J. Jeffcoate and K. G. Harding, "Diabetic foot ulcers," *The Lancet*, vol. 361, no. 9368, pp. 1545–1551, May 2003.
[3] S. R. Colberg *et al.*, "Exercise and type 2 diabetes: The American college of sports medicine and the American diabetes association: Joint position statement," *Diabetes Care*, vol. 33, no. 12, pp. e147–e167, Nov. 2010.
[4] S. Tesfaye, N. Chaturvedi, S. Eaton, J. Ward, C. Manes, C. Ionescu-Tirgoviste, D. Witte and J. Fuller, "Vascular Risk Factors and Diabetic Neuropathy", *New England Journal of Medicine*, vol. 352, no. 4, pp. 341-350, 2005.

Ultrasonic Sensors Height and BMI Device

Alexander Nguyen, Michael Heath, Anthony Messina, Jiang Wu, and Ying Sun*
Department of Electrical, Computer and Biomedical Engineering, University of Rhode Island

***Abstract*- This project is based on developing and creating a device that utilizes ultrasonic sensors to determine height. The device is intended to be used as an alternative to the other more bulky and expensive models found either at home or at your doctor's office. The device will be created for people to use it on their own with ease without requiring help. The objective is to make it both aesthetically pleasing, easy to use, and as cost effective as possible. The design includes two ultrasonic sensors, composite framework, a PIC controller, wiring, and C++ coding in its design.**

Keywords—Ultrasonic; Height Measuring; BMI

I. Introduction

Height and weight are important indicators of human health.. Most people likely know their height up to a certain point. But even being half an inch to an inch shorter could be a very important indicator to your health. For children, it is important to monitor that they are growing at a healthy rate, and for the elderly it is important to monitor whether or not their height decreases. This could be an indicator of osteoporosis. If this technology is paired with a scale, a person's Body Mass Index (BMI) can easily be calculated. Doctors commonly use BMI as important indicators for diabetes and heart disease.

Currently, methods to measure height and weight are archaic, take too much time, and usually required more than one person. These measurements, especially when directly linked to health, need to be as accurate as possible, and in a way that isn't time consuming. Ultrasonic sensors provide a solution to this problem. They work similar to echolocation in bats by sending out a sound wave and then determining the distance of something by listening for the return echo.. Based off of that information the device is able to determine distance to an object, which is in this case the floor. Having two ultrasonic sensors allows for two separate measurements, and by averaging both the design team was able to attain more accurate readings, while also compensation for non-level orientations.

This Project aims to make it easy for any person to take their measurements at home, while maintaining accuracy. When designing the prototype the team kept in mind how effective the device is, the cost of the materials, how easy it is to assemble and manufacture, and finally how consumer friendly it is. The team has done extensive research on the market for such an item, and has found quite a lucrative spot for such a device. Upon careful inspection there are a few products that are similar in theory, but in design could not be more different. These items cost anywhere from six hundred dollars to one thousand dollars. The design only cost around forty dollars to create. With this in mind the team has created a prototype that is pleasing to use for the consumer, but more importantly is accurate in determining height measurements.

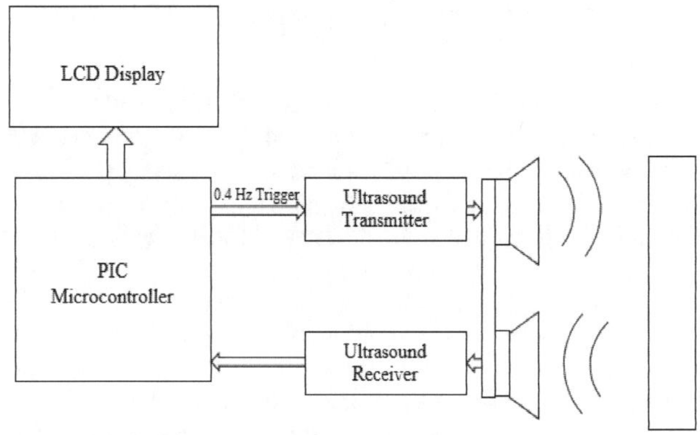

Figure 1. Block Diagram

II. Methods

A. Design Process

The design process began with the programing and implementation of the HC-SR04 ultrasonic sensor. This was first implemented onto a breadboard to first ensure that the idea would work before implementing it in the final design. The circuit includes a PIC18F4525 microcontroller, 4 MHz resonator, ultrasonic sensors, a 9 to 5 V converter, and one resistor. Initially before building the prototype the team needed to check to see if the ultrasonic sensors were functioning and how accurate they were. After this information was found the team began the design phase. Some things that the design team kept in mind while the prototype was created are the following:

1. The device should reach the ground, unobstructed by any obstacles.
2. The design should include two ultrasonic sensors at opposite sides so the average of the two can be taken.
3. The breadboard, wiring, and other components should be covered in a housing unit.
4. The prototype should be built so that it is kept stable at the top of your head with very minimal movement.

All of this was kept in mind when the team designed the functioning prototype.

B. The Prototype Build

The Software

Designing began with coding the PIC microcontroller using MPLab and C++ coding. The team programmed the microcontroller so that a trigger signal is sent to the ultrasound and that length of the return signal would be measured upon returning to give us a certain distance. This was recorded using a counter, which we then converted to feet, inches, and tenths of inches. The accuracy was improved with the second ultrasound, and a simple average command in the code to implement the addition. The information that was gained from the ultrasound was then coded to display on the LCD screen.

The Hardware

The hardware for our design included the PIC18F4525 microprocessor, a 4 MHz resonator, and 9 to 5 V converter, a 9 and 5 V battery, the two ultrasonic sensors, a resistor, the wiring, and the LCD to display the height. The final step was to build something that would house and encase the board and sensors. To do this wood composite board was cut so that the board and sensor can lay on them. Then two holes were cut at each end so the sensors could fit in there snugly. The breadboard lays on top of the wood composite plank in the center, and is encased in a housing of the same material. The LCD screen is mounted through a hole cut in the front of the housing, for easy viewing. The sensors are on each end of the plank to provide a large distance between them.

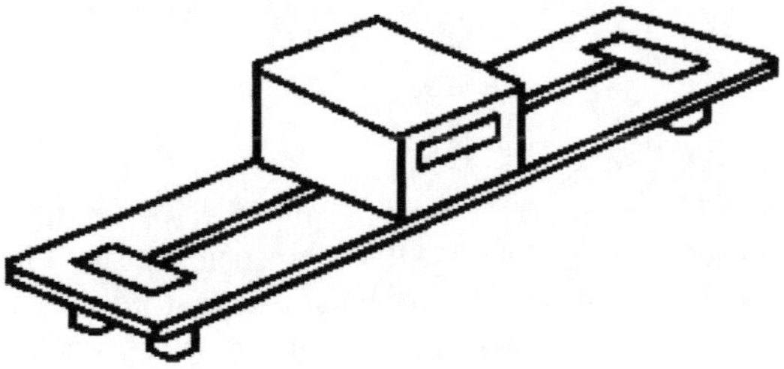

Figure 2. Sketch of basic prototype.

III. Results

Upon testing our prototype we found that it was functioning correctly and is very accurate. After measuring the same value 10 times, at most the average measurement was 0.4 inches off. The materials used to create our prototype are both cheap and lightweight. The device's weight is hardly noticeable when upon the subject's head and is very easy to hold in place.

IV. Discussion

This project was started this school year and developed in hopes of easing the process of measuring height, without sacrificing accuracy. A prototype was designed so a person alone at home could do this by themselves without spending a significant amount of money. The market for health and fitness is large and growing, and the need for new age products is very prevalent. It is the team's' belief that this device has great potential on such markets, with the only competitors being way too out of budget that only doctor's offices can afford them.

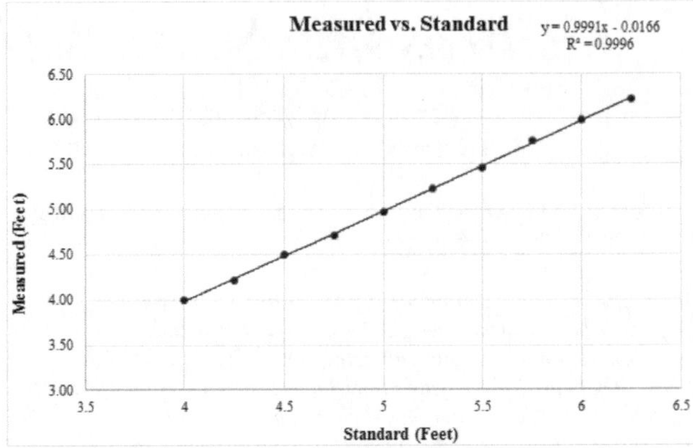

Figure 3. Graph of actual height compared to measured height.

Std	4	4.25	4.5	4.75	5	5.25	5.5	5.75	6	6.25
Avg	4	4.21	4.5	4.71	4.98	5.23	5.46	5.75	5.99	6.22
Std Dev	0.05	0.04	0.04	0.02	0.06	0.03	0.02	0.05	0.04	0.01
% E	0.10	0.84	0.11	0.77	0.48	0.48	0.76	0.06	0.21	0.47

Figure 4. Table of Actual height vs. the Average of 10 readings on our device, along with standard deviation and percent error.

In the future, we hope to add Bluetooth technology which will link the height measurement to an app and a Bluetooth scale. The app would be configured to determine a BMI calculations and notify you on your current health status. The group hopes in the future that they can make the design more easily manufacturable and aesthetically pleasing, so the user feels more comfortable using such device. Thus the device should be very marketable

References

[1] P. Cheng and Zhi Ping, "Portable Human Height Measuring Device" Patent US 6847586, Google patents and applications

[2] Sanoner, H. (2008). Distance measuring device for acoustically measuring distance. The Journal of the Acoustical society of America, 123(2), 583. doi:10.1121/1.2857655

[3] Ultrasonic Ranging Module HC – SR04. (n.d.). Retrieved from http://www.micropik.com/PDF/HCSR04.pdf

[4] Microchip PIC18F2525/2620/4525/4620. (n.d.) Retrieved from http://www.microchip.com/downloads/en/DeviceDoc/39626e.pdf

Image-based Open/Closed Eye Status Determination for Embedded System

John Paquet III, Andrew Rosenberg, Rory Makuch, Jiang Wu, Ph.D., Ying Sun, Ph.D.
Department of Electrical, Computer and Biomedical Engineering, University of Rhode Island
Kingston, Rhode Island, USA
Correspondence email: yingsun@uri.edu

***Abstract*—In this paper, we propose a simple SVM, or support vector machine, classifier-based algorithm for determining whether eyes are open or closed in images. Due to its computational simplicity, it can be implemented on embedded systems, such as the Blackfin Low-Power Imaging Platform (BLIP) by Analog Devices, Inc. This design provides an inexpensive and practical solution to drowsiness detection functionality in applications such as assistive devices for vehicle drivers or machinery operators.**

Keywords—eye status; image processing; machine learning; classifier; drowsiness detection; embedded systems

I. Introduction

Drowsiness detection plays an important role in advanced driver assistance systems (ADAS) in the latest vehicles by providing drivers fatigue information to avoid falling asleep while driving. Many alerting methods have been proposed based on a variety of information, including biomedical signals, driver behavior, stereo cameras, and IR cameras. Lately, low-cost video-based detection is receiving more and more attention [5]. Most of the reported research in this area requires computation-intensive eye/pupil shape recognition, such as the Haar Cascade Classifier and Camshift algorithms used to obtain eye blink detection [3]. In this paper, we propose a simple histogram-based SVM classifier for open/closed eye status determination targeted at less demanding computing platforms, such as the BLIP DSP board by Analog Devices.

The BLIP is a low-power, low-cost embedded computer vision board that is capable of many real-time video sensing applications, such as room occupancy determination and motion tracking. Images and videos captured by the BLIP camera can be analyzed in real time, and the USB ports on the board allow for external output video storage or display. Previous work at URI [2] with the BLIP includes using frame subtraction to monitor the activity of the elderly by identifying changes in consecutive images, thus determining whether an individual has moved and also to what degree the motion has occurred. In this research work, we continue using the same platform, but with the purpose of determining whether eyes are open or closed.

II. Methods

A. *Algorithm*

Open/closed eye status determination can be considered as a classification problem, and different machine learning classification methods can be utilized. They all share the same procedure: sample data collection, feature extraction, coefficient training, and prediction. For this particular research, after many experiments, histogram features and an SVM classifier are chosen.

Fig. 1 shows the process of feature extraction. The captured eye images are first converted to grayscale images if they are only available in RGB or YUV format, since the luminance components include the most important information about eye status. To eliminate the variation in image exposure, pixel-wise maximum/minimum normalization is applied to each grayscale image. In reality, if the exposure setting of the camera is fixed, this step can be omitted.

An 8-bin histogram is then computed across each grayscale image. The number of bins is chosen to be 8 so that the histogram can be easily computed by no more than 3-step shifting of the 8-bit pixel value while providing sufficient information for classification afterwards. The first 4 bins of the histogram are chosen to be the 4-dimensional feature vector. As a standard procedure, the feature vector is normalized to a standardized data set

$$f_{ij} = (h_{ij} - \mu_i)/\sigma_i, \quad i = 1,2,3,4 \text{ and } j = 0,1,\cdots N-1 \qquad (1)$$

with a mean of zero and a standard deviation of one, where h_{ij} is the ith feature of sample j and μ_i, σ_i are the mean and standard deviation, respectively, of the ith feature of the N samples. In order to train the classifier, eye images are captured and labeled as open or closed, and the collection of images is then split into a training set and a test set. The training set is used to find the classifier model coefficients, and the test set is used to assess the effectiveness of the model and coefficients.

A standard SVM classifier engine can be readily applied to the feature vectors of the captured sample images. The learned model coefficients are then used for model prediction of the real-time eye images to provide automatic eye status determination.

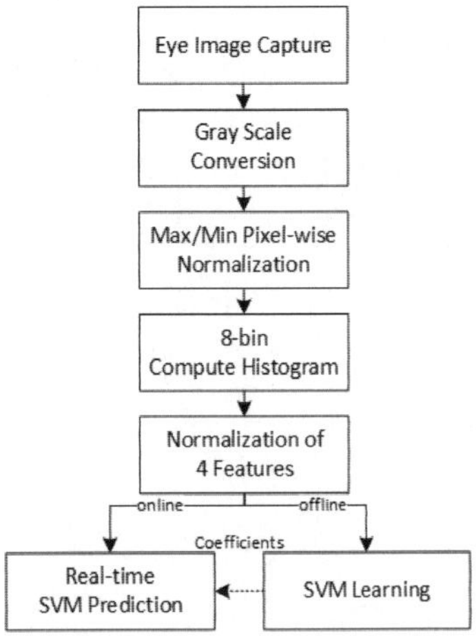

Fig. 1. Feature Extraction Diagram

B. Hardware

The BLIP board is used to capture eye images and will later perform real-time prediction. Thanks to its low primary voltage domains of 3.2, 1.8, and 1.1 volts and its petite 2.5-inch by 3.5-inch dimensions, this embedded platform can be easily integrated into many systems. To access the on-board Aptina ASX340AT imaging sensor, which captures 640-pixel by 480-pixel images at 30 frames per second, AD Vision Sensor Controller 1.0.1 is installed on a PC running several other Analog Devices programs for development and demonstration purposes.

III. RESULTS

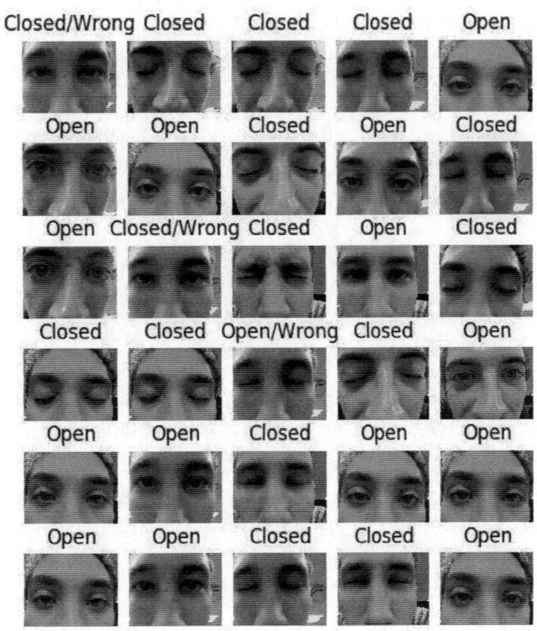

Fig. 2. 30 grayscale images tested against the machine learning algorithm. The eye status of 27 of the 30 images was successfully determined.

The built-in camera of the BLIP board was used for sample eye image collection. The images are labeled and processed by a Python script for normalization and feature extraction. Then, the SVM model provided by the Python machine learning package Scikit-learn is used for model training and verification. In our test, 152 images of three different individuals with open and closed eyes were captured for coefficient learning: 122 for the training set and 30 for the test set. Fig. 2 shows the classification results for the test images. A detection rate of 90% or better is achieved.

IV. DISCUSSION

The next major step of the project is to program the BLIP microprocessor. The BLIP embedded system will have an SVM prediction function implemented and executed on real-time eye images, based on the already-learned model coefficients. The code for the embedded system will be developed in C with CrossCore Embedded Studio 2.1.0, an integrated development environment made by Analog Devices. Also, the image populations to train the machine learning algorithm and to test it should be extended and further diversified to increase the algorithm's robustness and further verify its functionality and efficacy. Future work includes using the system to alert tired drivers through determining if the eyes have been closed for a duration longer than that of a typical blink or if the rate of blinking is below normal. In addition, the BLIP system, by differentiating intentional eye blinking from involuntary blinking, could allow for easier control of devices for people suffering from spinal cord injuries.

ACKNOWLEDGMENT

The authors would like to thank Analog Devices, Inc. for the donation of the BLIP board for this project.

REFERENCES

[1] Analog Devices, Inc., ADZS-BF707-BLIP2 Board Evaluation System Manual, Revision 1.0, April 2015.

[2] J. McVaney, V. Danielson, S. Ramos, J. Wu, and Y. Sun, "An Activity Analyzing System for Elders Based on Standalone Embedded Image Platform," 2016, unpublished.

[3] A. A. Mohammed and S. A. Anwer, "Efficient Eye Blink Detection Method for disabled-helping domain," International Journal of Advanced Computer Science and Applications (IJACSA), Vol. 5, No. 5, 2014.

[4] A. Soetedjo, "Eye Detection Based-on Color and Shape Features," International Journal of Advanced Computer Science and Applications (IJACSA), Vol. 3, No. 5, 2011.

[5] I. García, S. Bronte, L. M. Bergasa, N. Hernandez, B. Delgado, and M. Sevillano, "Vision-based drowsiness detector for a Realistic Driving Simulator," 13th International IEEE Annual Conference on Intelligent Transportation Systems, 2010.

GUI Based Optic Disc and Cup Characterization from Fundus Images

Samuel Spink, Mitchel Apatow, Scott Goyette, Ying Sun, Ph.D., and Jiang Wu, Ph.D.
Biomedical Engineering, University of Rhode Island, 4 East Alumni Avenue, Kingston, RI 02881

***Abstract*—The aims of this project were twofold: (1) to develop an accurate, yet simple and efficient algorithm that estimates the optic cup-to-disc ratio from input retinal images (a critical feature in the evaluation of glaucoma), and (2) to implement this algorithm in a graphical user interface (GUI) that can aid optometrists and other trained professionals in this characterization process. The first aim addresses the possibility of having this diagnostic tool implemented in devices that can be transported to areas currently devoid of appropriate medical care at low prices. Both goals have been completed successfully. The developed algorithm works as follows: by selecting the center point of the optic disc and optic cup on a retinal image, radial lines are projected outward in a circular pattern that detect the edges by determining the greatest change in pixel intensity. The points of greatest difference are connected to form polygonal approximations of the desired features. The areas of the polygons are calculated and the ratio between the two areas is determined, providing a reliable estimate to the relevant cup-to-disc ratio.**

Keywords—glaucoma detection, optic cup, optic disc, graphical user interface, GUI, edge detection, radial lines, dead zone.

I. Introduction

Glaucoma affects 60 million people worldwide, and is the leading cause of vision loss in the United States. [1] Glaucoma is a disease that damages the eye's optic nerves due to increased intraocular pressure. [2] Increased intraocular pressure results from partially or fully blocked drainage canals located around the iris, which prevent the aqueous humor from draining appropriately. When intraocular pressure increases, a force is asserted on the optic nerve, which restricts blood flow through the blood vessels. This can result in permanent, irreversible damage. [2] Symptoms of glaucoma include nyctalopia, peripheral vision loss, light sensitivity and blurred vision. [2] If glaucoma is detected at an early stage, the symptoms can be mitigated through medication or surgery. [2]

One characteristic that is associated with the presence of glaucoma is an increased optic cup-to-disc ratio. [3] The optic cup is the central portion of the optic disc that lacks nerve fibers. When glaucoma progresses to a severe state, the area of the cup increases and approaches the outer edge of the optic disc. Therefore, as the cup-to-disc ratio increases, the likelihood of glaucoma increases. It is desirable to develop accurate, efficient, and simple methods of characterizing this feature in order to make the diagnostic procedure easy and affordable to implement in medical practice.

II. Methods

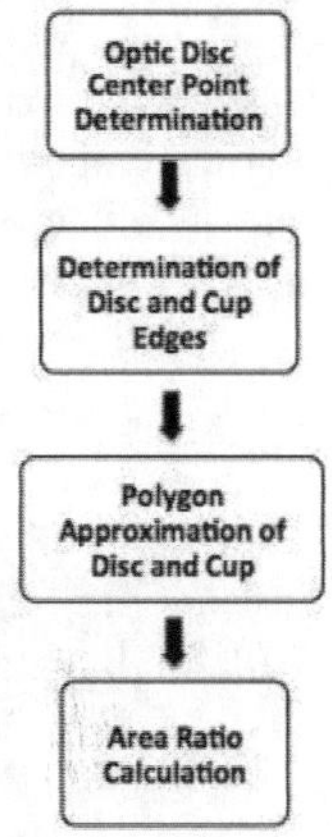

Figure 1: Flow chart of algorithm

Figure 1 displays the general procedure for performing this disc and cup characterization. The first step in the process is the determination of a center point from which radial lines will be launched. This is done in our GUI by manually clicking a center point based on visual inspection of the fundus image.

From this point, the edges of the cup and disc are detected using an algorithm that tracks pixel intensities along the radial lines projected outward from the designated center point. Each line consists of a discrete number of intensities sampled at a certain step size along the radial line. [4] This sampling does not actually begin until a specified "dead zone" has been cleared, which is a predefined circular region in which no samples are taken in order to avoid mistaking blood vessels for disc and cup edges (the vessels converge right near the center of the optic disc). The sampling locations are described by the following formulas:

$$x(j) = x_c + d * \cos(\theta) + p * (j-1) * \cos(\theta) \quad (1)$$

$$y(j) = y_c + d * \sin(\theta) + p * (j-1) * \sin(\theta) \quad (2)$$

In equations (1) and (2), x_c and y_c refer to the x and y pixel coordinates of the center point, $x(j)$ and $y(j)$ refer to the x and y coordinates of sample j along a radial line projected at angle θ to the horizontal. The variables d and p refer to the dead zone radius and step size, respectively.

For most of the lines, the sample locations do not fall on an integer pixel location, and thus the average of the four surrounding pixels is used to interpolate the intensity at these points. The change in intensity is determined between each consecutive pair of pixels along a line, and the point at which the greatest derivative is found is deemed an edge point of the disc. The derivative used is a simple discrete derivative shown here:

$$D(i,j-1) = M(i,j-1) - M(i,j) \quad (3)$$

In equation (3), $D(i,j-1)$ refers to the derivative D at the previous step $(j-1)$ along radial line i. The term $M(i,j-1)$ is the interpolated pixel intensity at the previous step on radial line i, and $M(i,j)$ is the interpolated pixel intensity at the current step.

The same process is then repeated to find the cup edges, except the lines only travel to the disc edge this time so that the next highest derivative is located. This means the success of the cup characterization is highly dependent on the success of the disc characterization. In an attempt to eliminate outlier points, the mean distance between the center point and disc edge points is calculated and compared to each of the individual distances. The same process is repeated for the cup characterization. Some upper and lower tolerance is allowed, while points whose distances lie outside this allowed range are replaced by a point of median distance. This median filtering of outliers prevents any one outlier from greatly skewing the characterization.

Once the edges of the optic cup and disc are located using the radial line technique, the edge points are directly connected, creating two polygons. The areas of the two polygons are calculated and a ratio is determined.

We developed this algorithm in Matlab, and presently a fully functional GUI exists in which one can adjust the various parameters (such as the dead zone radius) and view the updated edge point detection and area ratio calculation live.

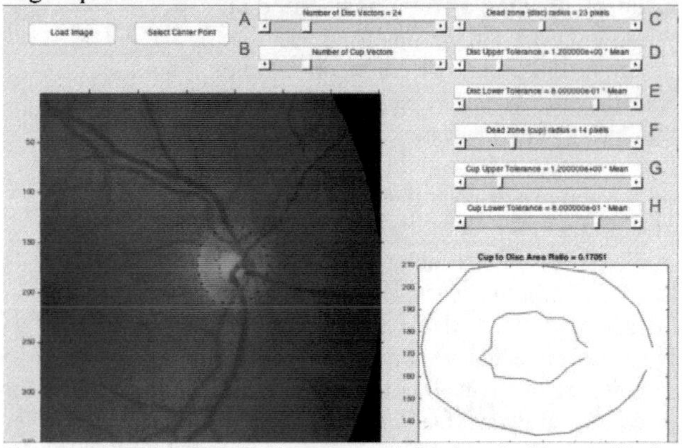

Figure 3: MATLAB GUI; The image on the left is the retinal image with disc (blue) and cup (purple) points shown. The plot on the bottom right depicts the polygonal approximations and cup to disc ratio (displayed above it). The "Load Image" and "Select Center Point" buttons allow a retinal image to be loaded, resized to focus on the optic disc, and can be clicked to select the center point. The 8 slider bars allow the following to be adjusted: A) The number of radial lines used for the disc edge detection B) The same parameter described in A, but for the cup detection. C) The dead zone radius for disc detection described in text. D) The upper tolerance of allowed distances between center and disc. E) The lower tolerance. F, G, H) Same parameters described in C, D, and E, but for cup detection instead of disc.

III. Results

The described algorithm can successfully locate the disc and cup edges on a number of retinal images. However, the values of some of the parameters can greatly influence the effectiveness of the algorithm, which makes the implementation of this algorithm in the GUI very useful. The ability to actively adjust various parameters described in Figure 2 allows one to see how each variable affects the characterization algorithm, and in turn make appropriate adjustments that improve the characterization for any particular image. Figure 3 demonstrates this usefulness by showing the results of using three different values for the dead zone radius during optic disc detection. In this figure the optic cup points have been suppressed for clarity. For this image, when the dead zone radius is set to 15 pixels, the optic disc points (circular points in figure 3) undershoot the apparent visual disc perimeter. When it is increased to 45 pixels, there is an overshoot (triangular points). A decrease in the dead zone to 30 pixels results in a very accurate outline of the optic disc (arrow shaped points). The algorithm reruns each time a slider is adjusted. The result of any adjustments can be seen and evaluated in real time which increases the efficiency of parameter optimization

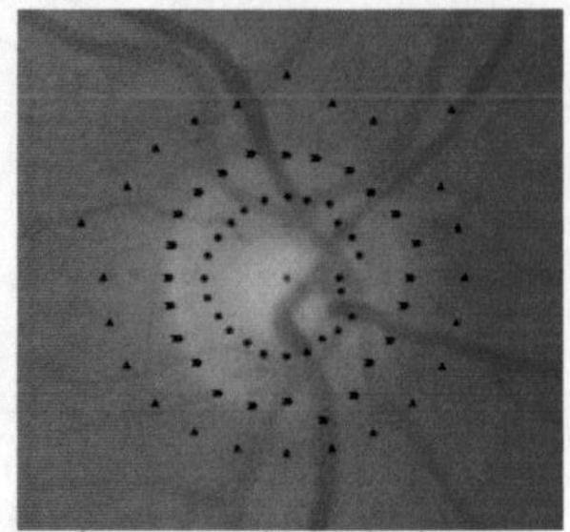

Figure 2: Each set of points shows the disc points at a particular dead zone radius setting; triangle set-45 pixels, arrow set-30 pixels, circle set-15 pixels.

IV. Discussions

The current functionality of this algorithm in Matlab shows promise for the evaluation of glaucoma on a variety of platforms. However, there are many improvements that can be made in future development. As mentioned previously, it would be desirable to have this same algorithm implemented in a cheap medical device that could be produced at low costs and potentially be used in developing parts of the world where medical care is subpar. In particular, the Blackfin Low Power Imaging Platform Board (BLIP) is one affordable option from Analog Devices that can run the developed algorithm. Additionally, the need for these special "dead zone" parameters could be removed by developing a method of automatically detecting and either removing or ignoring blood vessel disturbances. One possibility we have started to examine is the identification of pairs of derivative peaks of opposite signs, a phenomenon typically indicative of blood vessel edges. This implementation would make the algorithm more robust since the regions that should be ignored would be tailored to the blood vessel geometries of each image. The GUI could also be used to further investigate correlations between cup-to-disc ratio and the presence of glaucoma. With this tool, it would be very easy and efficient to get ratio estimates for thousands of known sample images, and use this data to more accurately determine risk indices for glaucoma based on the cup-to-disc ratio.

V. Acknowledgment

The authors would like to acknowledge the grant support from the 2016-2017 Undergraduate Research Initiative Awards program of the University of Rhode Island.

VI. References

[1] Weinreb, Robert N., MD. "Glaucoma Worldwide: A Growing Concern." Glaucoma Worldwide: A Growing Concern | Glaucoma Research Foundation. Gleams, 15 Dec. 2010. Web. 28 Nov. 2016.

[2] Chudler, Eric H., Ph.D. "Eye Diseases - Glaucoma." Eye Diseases - Glaucoma. University of Washington, n.d. Web. 28 Nov. 2016.

[3] Hoffmann, Esther M. et al. "Optic Disk Size and Glaucoma." Survey of Ophthalmology. U.S. National Library of Medicine, 2007. Web. ne 28 Nov. 2016.G. Eason, B. Noble, and I. N. Sneddon, "On certain integrals of Lipschitz-Hankel type involving products of Bessel functions," Phil. Trans. Roy. Soc. London, vol. A247, pp. 529–551, April 1955. (references)

[4] Guo, Yu et. al. "Automated Detection and Delineation of Mitochondria in Electron Micrographs of Human Skeletal Muscles." Microscopy Research and Technique 139th ser. 63.133 (2004): n. pag. .